LEV PONTRYAGIN

GENERALIZATION OF NUMBERS

TRANSLATED FROM RUSSIAN BY
NIKOLAY POLYAKOV AND PAUL EDWARDS

2010

Foreword

The name of Lev Semenovich Pontryagin is well known in mathematics. He made fundamental contributions to the field, especially in topology and control theory. For Russian students of mathematics he is known in the first place as an author of excellent textbooks. The textbooks "Ordinary Differential Equations", and "Topological Groups" are written with unmatched lucidity, something bordered on magic. They became classical works that taught and inspired many generations of Russian mathematicians.

There was a reform of high school mathematical education in Russia in the Nineteen-Seventies. After a while it became apparent that this reform had serious problems. Pontryagin got involved in a fight to reverse this reform. He believed that the approach taken by the authors of the reform made even simple topics look abstruse, and in general alienated students from mathematics. To alleviate the effects of the reform and to lure high school students back into mathematics, he wrote a series of books that show what mathematics is really about. This book is one of those.

The goal of this book is to show its readers that some truly deep and profound mathematical developments are in fact quite approachable for them. Pontryagin did not use the "Scientific America" style for presentation of mathematics by not using any complicated formulas, a usual approach for a book aimed at young readers. Instead he surprisingly took a very different, much more ambitious approach. He decided to use all the complex formulas the subject needs and to write the book in such a way that the reader would be able to understand all these formulas. Pontryagin, with his exceptional clarity of presentation, was perhaps the only mathematician up to this task.

Even so, this book is not an easy read. It requires certain effort from its readers, who will be rewarded for this effort not only with an understanding of some major mathematical developments but also with the acquisition of tools and knowledge that will allow them to go further and deeper into the study of numbers.

The subject of this book is about the possible generalizations of the notion of numbers. The high point of the book is a theorem that Pontryagin proved when he was young and worked on some topology problems under the mentoring of Aleksandrov. It was his first significant result and marked the beginning of his outstanding mathematical career.

When the translators talked to Pontryagin's widow she told us that the goal of this book was to reach out to young people curious to know the nature of things and to spread the word of how beautiful and deep mathematics can be. There are no borders in science, but there are barriers to learning caused by language. To overcome them we translated this book into English.

From his vantage point in the first half of the 20th Century, Pontryagin believed that quaternions could only play a minor role in mathematics. However, quaternions have had a revival in the late 20th century. They proved useful in describing rotations in space, something so very well covered by Pontryagin in this book. Rotations done by using quaternions are more compact and faster to compute than by using matrices. Therefore, quaternions are used in computer applications like graphics, robotics, and signal processing, in sciences like physics and bioinformatics, and in applications like molecular dynamics and orbital trajectories.

The other extension of numbers not covered in this book is the Cayley Numbers, or octonions. Because they don't offer associative multiplication, the octonions receive less attention than the quaternions. Nevertheless, they are related to a number of interesting structures in mathematics, such as in the exceptional Lie groups. Additionally, octonions have applications in string theory, special relativity, and quantum logic.

We would like to thank Aleksandr Polyakov for his numerous contributions to the translation, valuable discussions, help with graphics design, and constant support.

We are very happy to acknowledge Dr. Iaroslav Kryliouk, our reviewer, who made many excellent suggestions and corrections that helped to greatly improve the quality of this book. Dr. Kryliouk has his BS and PhD from Moscow University. He is currently a full-time instructor of Mathematics at De Anza Community College, Cupertino, California.

We would also like to thank Svetlana Polyakova for her help with design of the cover of this book.

Nikolay Polyakov   Campbell
Paul Edwards   June 2010

Contents

i

The notion of a number in mathematics was formed gradually over a long period of time under the influence of experience and the needs of mathematics itself. Finally the notion of the real numbers, which we assume in this book to be well known to the reader, was formed.

The development of the idea of a number did not stop at the real numbers. The internal needs of mathematics caused the creation of the complex numbers. Today, the theory of functions of complex variables, that grew out of the base of the complex numbers, has vast practical applications. Much of this book is devoted to the complex numbers. The proof of the *Fundamental Theorem of Algebra*, that every polynomial has at least one real or complex root, is presented here. Then the division of a polynomial by linear polynomial divisors, which depends on the Fundamental theorem, is studied in great detail. At the same time, the division of polynomials by each other and Euclid's algorithm are used as supplemental tools.

Because the complex numbers proved to be extremely important and useful in mathematics, mathematicians attempted to further generalize the idea of number. Thus appeared quaternions, but at the price of sacrificing the commutativity of multiplication. Because of this absence of commutativity of multiplication, it turned out to be impossible to build a theory of functions of quaternion variables. As a consequence, quaternions found very little application in mathematics. Quaternions are useful in describing revolutions of three and four dimensional Euclid spaces, but this is incomparable in significance with applications of the complex numbers. A description of quaternions and their application to studying revolutions of three and four dimensional Euclidean spaces is presented in this book. That section is concluded with the proof of the Frobenius theorem, which states that any further development of the notion of a number in the direction of quaternions is impossible.

The development from rational numbers to real numbers happened due to the internal logic of the development of mathematics, rather than by need of applications, since it is possible to make any measurement with an arbitrary precision using only rational numbers. The idea of a real number was brought into existence by a mathematical discovery, a consequence of the Pythagorean theorem, that a diagonal of a square with side of length 1 cannot be precisely measured using rational numbers alone. It is like the real numbers fill in the gaps between the rational numbers, which leads to the Cauchy criterion being not only a necessary but also a sufficient condition for the convergence of an infinite sequence. This is a crucial fact in mathematics. The real numbers represent a continuous substance that the rational numbers are submerged in. Here it becomes absolutely clear that not only the existence of the operations of addition, multiplication, subtraction, and division are intrinsic to numbers, but equally the notion of a limit of a sequence. That is, the notion that a particular number is a limit of a sequence as the sequence index tends to ∞ is well defined.

The set of objects with defined addition, multiplication, subtraction, and division operations, together with the definition of limit, is a natural logical generalization of the notion of a number. It turns out that there are not at all that many such generalizations. It is precisely their description to which this book is dedicated.

The move from rational to real numbers is based on the notion of very small rational numbers. It turns out, that in addition to the completely natural notion of rational number smallness, there is a different notion associated with some prime number p. Connected to this notion of smallness, the expansion of rational numbers leads to the creation of the p-adic numbers. Today such numbers have important applications in number theory and are described in this book.

The objects for which algebraic operations are possible are so-called modulo p numbers. Rational functions of t that have coefficients in terms of modulo p numbers constitute a set with possible operations of addition, multiplication, subtraction, and division, and also, naturally, the notion of smallness. Expanding this system of rational functions so that the resulting set of object is closed, from the point of view of limit, or equivalently, that the Cauchy criterion is necessary and sufficient for convergence, we come to an examination of the infinite series (sequences?) of t. In the end of this book, Kovalsky's theorem which describes to some degree any system of numbers that contains algebraic operations and the limit is presented.

This book is devoted to a description of the sets of objects with algebraic operations and limit defined, that represent the logically plausible generalizations of numbers. Applying some very simple and natural constraints on such a set we come to the conclusion that there are no logical possibilities acceptable for mathematics that are analogous to the real and complex numbers, except for the real and complex numbers themselves. This shows that the real and the complex numbers appear in mathematics by no accident, but as the only logically possible objects with qualities that we naturally expect from numbers.

In conclusion, I want to thank S. M. Aseev for great help in the editing of this book.

CHAPTER 1

Complex numbers

In this chapter I briefly describe how the complex numbers came about and gradually gained ground in mathematics. Then I give a definition of the complex numbers, the operations on them, and their geometric interpretation. On the way I prove formulas using sine and cosine of the sum of two numbers that are closely connected with the operation of multiplication of complex numbers.

1.1 A Brief History of Complex Numbers

From the study of mathematics it is well known that the negative numbers were introduced, in the first place, in order to make possible the subtraction operation, which is the inverse operation to addition. It was for a similar reason that the complex numbers appeared in mathematics. If we consider only the real numbers, then it is not always possible to take the square root of a number, the inverse operation of taking the square of a number, because it is not possible to take the square root of a negative number. This problem is not great enough reason for introducing new numbers in mathematics though. It turns out that performing calculations following common rules on the expressions containing the square root of a negative number, it is possible to obtain a result that does not contain a square root of a negative number. In the Sixteenth Century Cardano found a formula for solving the cubic equation.

It turns out that particularly in the case when all three roots of the equation are real his formula contains a square root of a negative number. In this way, it became evident that by performing calculations on expressions containing the square root of a negative number it is possible to obtain reasonable results. For this reason, these kinds of square roots started to be used in mathematics. These numbers were named *Imaginary Numbers*, and in this way they gained the right to an illegal existence. Full "citizen" rights were given to these numbers by Gauss at the turn of the Eighteenth Century.

He named them *the complex numbers*, gave a geometric interpretation of them, and what is most important, proved the *Fundamental Theorem of Algebra* that states that every polynomial has at least one real or complex root.

1.2 Definition of Complex Numbers

We assume that we familiar with the real numbers. We know that there are addition and multiplication operations as well as subtraction and division operations, which are the inverse operations for the first pair. The rules that govern these operations are well known, we use them almost automatically, and I am not going to spend time here on giving definitions for them. The set of objects for which the operations of addition, multiplication, subtraction, and division are defined, and for which the same rules apply as for the operations on real numbers, is called in modern abstract algebra a *field*.

Therefore, from the modern abstract algebra point of view, the set of all real numbers, D is a field. Let's consider the problem of extending the notion of number, or, in terms of modern algebra, let's consider the problem of extending the field D to the field K^2, so that in this new field K^2, the equation

$$z^2 + 1 = 0$$

has a solution. Let's name the element of the set K^2 that satisfies this equation, i.e. is the root of this equation, by i. Thus, for i the following is true:

$$i^2 = -1 \tag{1.1}$$

The field K^2 contains all the real numbers, and i, and since the operations of multiplication and addition are possible there, K^2 should contain all polynomials of i with real coefficients, and, in particular, all polynomials of first degree, i. e. the expressions of the following type:

$$z = x + yi = x + iy$$

where x and y are real numbers. Expressions of this type are called *complex numbers*. We define operations over them as operations on polynomials, taking into account condition (1.1). Complex numbers of the type

$$z = x + 0i = x$$

are called real numbers. Complex numbers of the type

$$z = 0 + yi = yi$$

are called pure imaginary numbers.

Let us suppose that $z_1 = x_1 + y_1 i$, and $z_2 = x_2 + y_2 i$ are two complex numbers. According to the previously stated rule the sum and multiplication of these complex numbers will be defined by the following equations respectively:

$$\begin{aligned} z_1 + z_2 &= (x_1 + y_1 i) + (x_2 + y_2 i) = \\ &= (x_1 + x_2) + (y_1 + y_2)i \end{aligned} \tag{1.2}$$

$$\begin{aligned} z_1 z_2 &= (x_1 + y_1 i)(x_2 + y_2 i) = \\ &= x_1 x_2 + (x_1 y_2 + y_1 x_2)i + y_1 y_2 i^2 = \\ &= (x_1 x_2 - y_1 y_2) + (x_1 y_2 + y_1 x_2)i. \end{aligned} \tag{1.3}$$

Note that to get the last equality we used (1.1).

If the number $z_1 = x_1$ was real, we would have gotten

$$x_1 z_2 = x_1 x_2 + x_1 y_2 i \tag{1.4}$$

From the formulas (1.2) and (1.3) it is clear that the the sum as well as the product of two complex numbers are complex numbers.

In order to make sure that the subtraction operation, the inverse operation to the addition operation, exists it is sufficient to find a number $-z$ which is the inverse to z. Similarly to check that the division operation exists, for any $z \neq 0$, it is suffice to find a number z^{-1} that is the inverse to the number $z = x + yi$. These numbers, as it is easy to see, are:

$$-z = -x - yi, \quad \text{and} \quad z^{-1} = \frac{x}{x^2 + y^2} - \frac{y}{x^2 + y^2} i$$

Therefore, the number z^{-1}, inverse to the given z, always exists when $z \neq 0$.

1.3 Geometric Representation of Complex Numbers

Let's choose some Cartesian coordinates on a plane P (Figure 1.1.) Let's place the complex number $z = x + iy$ at the point z with coordinates (x, y). Let's also label the vector going from the origin to the point z, by z. Thus, z denotes now at the same time a complex number, a point on the plane that represents this complex number, and the vector corresponding to this complex number. In this representation all real numbers are located on the abscissa axis. This is the reason why the abscissa axis is called *the real axis* of P the plane of the complex variable. The pure imaginary numbers are located on the ordinate axis, that is called *the imaginary axis* of P the plane of the complex variable. Zero is located at the origin of the coordinate system (see Figure 1.1.)

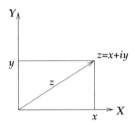

Figure 1.1: Geometric representation of a complex number.

The length of the vector z is called the *modulus* of the complex number $z = x + iy$ and is denoted by $\mid z \mid$:

$$\mid z \mid = +\sqrt{x^2 + y^2}$$

The complex numbers that satisfy the condition $\mid z \mid = 1$ lay on the circle with radius 1 with the center at the origin of the coordinate system. The number 1 is also located on this circle. Let's draw out from the point 1 along the circle a segment of length φ in the counterclockwise direction. Let's denote the end of this segment by (φ). If φ is negative, we should draw out the segment of the length $\mid \varphi \mid$ from the point 1 in the clockwise direction. It is known that the abscissa of the point (φ) is called $\cos \varphi$, and the ordinate, $\sin \varphi$. Thus the complex number (φ) is represented by the following formula:

$$(\varphi) = \cos \varphi + i \sin \varphi \qquad (1.5)$$

So, each complex number with modulus equal to 1 is represented by the formula (1.5). If z is an arbitrary complex number, with modulus $\mid z \mid = \rho$ not equal to 0, then the number z/ρ is the complex number with modulus equal to 1, and therefore, may be represented by formula (1.5). From the equality

$$z/\rho = \cos \varphi + i \sin \varphi$$

we get

$$z = \rho(\cos\varphi + i\sin\varphi). \qquad (1.6)$$

Formula (1.6) is called the *trigonometric form* of a complex number. The number φ is called the *argument* of the complex number. If the modulus of the complex number z is not equal to 0, the argument is defined up to the addend $2k\pi$, where k is an integer number. If the modulus of ρ is equal to 0, formulas (1.6) also makes sense, but in that case the argument of the complex number is not defined.

Numbers ρ, and φ are called the *polar coordinates* of the point z.

Let us consider now a geometric interpretation of the operations on the complex numbers.

From the formulas (1.2) and (1.4), it follows that the addition and multiplication operations on the complex numbers are exactly the same as those operations for vectors.

The geometric interpretation of the addition operation on the complex numbers is obvious: vector $z_1 + z_2$ is a diagonal of the parallelogram built on the vectors z_1 and z_2. From this fact follows the important inequality:

$$\mid z_1 + z_2 \mid \leq \mid z_1 \mid + \mid z_2 \mid \qquad (1.7)$$

In order to give a geometric interpretation of the complex number multiplication operation, we apply the operation of rotation of a vector, or what is the same, of a complex number. Rotating some vector z in the counterclockwise direction about the angle α, we get some new vector $R_\alpha(z)$. From a geometric consideration, it is clear that the operation R_α, rotating about the angle α, has the following properties. If a is a real number, then

$$R_{\alpha+\beta}(z) = R_\alpha(R_\beta(z)),$$
$$R_\alpha(az) = aR_\alpha(z),$$
$$R_\alpha(z_1 + z_2) = R_\alpha(z_1) + R_\alpha(z_2)$$

From the two last equations it follows that if a_1, and a_2 are two real numbers, then

$$R_\alpha(a_1 z_1 + a_2 z_2) = a_1 R_\alpha(z_1) + a_2 R_\alpha(z_2) \qquad (1.8)$$

It's also obvious that

$$R_\alpha(1) = \cos\alpha + i\sin\alpha \qquad (1.9)$$

Let us prove now that rotation of the complex number $z = x + yi$ about the angle α is equivalent to multiplication of this number by the complex number $\cos\alpha + i\sin\alpha$, in other words, let's prove the following formula for the operation R:

$$R_\alpha(z) = (\cos\alpha + i\sin\alpha)z. \qquad (1.10)$$

Let us consider first rotation about the angle $d = \pi/2$. In that case $\cos d + i\sin d = i$, and formula (1.10) transforms into $R_d(z) = iz$. Geometrically it's obvious that $R_d(1) = i, R_d(i) = -1$. On the other hand, $i \cdot 1 = i$, and $i \cdot i = -1$. Therefore,

$$R_d(1) = i \cdot 1, R_d(i) = i \cdot i$$

From formula (1.8) it immediately follows that:

$$iz = i \cdot (x + iy) = x \cdot i + y(-1) = xR_d(1) + yR_d(i) =$$
$$= R_d(x \cdot 1 + y \cdot i) = R_d(x + iy) = R_d(z).$$

We proved formula (1.10) for $d = \pi/2$.

Let us take some real number α. For $\hat{z} = \cos\alpha + i\sin\alpha$ we get

7

$$\hat{z} \cdot i = i\hat{z} = R_d(\hat{z}) = R_d(\cos\alpha + i\sin\alpha) =$$
$$= \cos(\alpha + \frac{\pi}{2}) + i\sin(\alpha + \frac{\pi}{2}) = \qquad (1.11)$$
$$= R_\alpha(\cos\frac{\pi}{2} + i\sin\frac{\pi}{2}) = R_\alpha(i).$$

Therefore, formula (1.10) is proven for $z = i$.

Now let us prove formula (1.10) for any arbitrary complex number

$$z = x + iy$$

From the formulas (1.8), (1.9), and (1.11) it follows that

$$R_\alpha(z) = R_\alpha(x + iy) = xR_\alpha(1) + yR_\alpha(i) =$$
$$= x(\cos\alpha + i\sin\alpha) + y(\cos\alpha + i\sin\alpha)i =$$
$$= (\cos\alpha + i\sin\alpha)(x + yi) =$$
$$= (\cos\alpha + i\sin\alpha)z.$$

Thus the proof of formula (1.10) is completed. ∎

Applying formula (1.10) to the complex number $z = \cos\beta + i\sin\beta$ we get

$$(\cos\alpha + i\sin\alpha)(\cos\beta + i\sin\beta) =$$
$$= R_\alpha(\cos\beta + i\sin\beta) = R_\alpha(R_\beta(1)) =$$
$$= R_{\alpha+\beta}(1) = \cos(\alpha + \beta) + i\sin(\alpha + \beta).$$

So

$$(\cos\alpha + i\sin\alpha)(\cos\beta + i\sin\beta) =$$
$$\cos(\alpha + \beta) + i\sin(\alpha + \beta)$$

Multiplying the complex numbers in the left part of the equation, using formula (1.3) we get

$$(\cos\alpha + i\sin\alpha)(\cos\beta + i\sin\beta) = (\cos\alpha\cos\beta - \sin\alpha\sin\beta) +$$
$$+ i(\sin\alpha\cos\beta + \cos\alpha\sin\beta)$$

That means that we got formulas for cosines and sines of the sum of two angles:

$$\cos(\alpha + \beta) = \cos \alpha \cos \beta - \sin \alpha \sin \beta$$
$$\sin(\alpha + \beta) = \sin \alpha \cos \beta + \cos \alpha \sin \beta$$

For any two arbitrary complex numbers that we can write in the form $\rho(\cos \alpha + i \sin \alpha)$ and $s(\cos \beta + i \sin \beta)$, we get:

$$r(\cos \alpha + i \sin \alpha) \cdot s(\cos \beta + i \sin \beta) =$$
$$= rs[\cos(\alpha + \beta) + i \sin(\alpha + \beta)].$$

Therefore, when two complex numbers are multiplied, their moduli are also multiplied, and their arguments are added.

Formula (1.12) may be obviously extended to any number of factors. If all these factors are equal, and equal, say to $r(\cos \alpha + i \sin \alpha)$, we get

$$[r(\cos \alpha + i \sin \alpha)]^n = r^n(\cos n\alpha + i \sin n\alpha)$$

This is a very interesting formula. It shows the way to get an nth root of an arbitrary complex number $\rho(\cos \varphi + i \sin \varphi)$. It turns out that the number of nth roots of any nonzero complex number, $\rho(\cos \varphi + i \sin \varphi)$, is equal to n, and all these roots are located on the circle with radius $\sqrt[n]{\rho}$ and center at the coordinate origin, and they are the vertices of the regular n-polygon inscribed in the circle. I leave this statement for the reader to prove.

In particular, the nth root of 1 has n roots that are vertices of the regular n-polygon inscribed in the circle with radius equal to 1 (see Figure 1.2 [1]) This can be written as a formula in the following way:

$$\sqrt[n]{1} = \cos \frac{2k\pi}{n} + i \sin \frac{2k\pi}{n}, \ k = 0, 1, 2, ..., n - 1.$$

---

[1]Figure 1.2 shows the roots of $\sqrt[7]{1}$, which are located at the points $\cos \frac{2\pi k}{7} + i \sin \frac{2\pi k}{7}$ for k = 0,1,...,6.

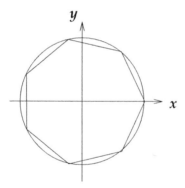

Figure 1.2: 7th root of 1.

Chapter 2

The Fundamental Theorem of Algebra

The proof of the *Fundamental Theorem of Algebra*, that any polynomial with complex coefficients has at least one complex root, is presented here. Throughout, the real numbers are considered a special case of the complex numbers. This theorem was first proven by Gauss in 1799 for a special case of a polynomial with all real coefficients. Gauss showed that any such polynomial has at least one real or complex root. From the point of view of modern abstract algebra, this theorem shows that the field of complex numbers is algebraically closed. This means that while investigating the roots of the algebraic equations (the roots of the polynomials) we cannot obtain any new types of numbers.

In this sense, the field of the complex numbers is radically different from the field of real numbers, which is not algebraically closed. At the same time, we should note that the complex number field is derived from the real number field by the mere addition of a single root of the equation

$$z^2 + 1 = 0$$

The proof of the *Fundamental Theorem of Algebra* is based not on the ideas of abstract algebra, but rather on the specific examination of the complex number field. The rigorous proof must be based on the precise definition of the real number and the precise definition of the continuity of functions. I describe here not the rigorous, but instead a geometrically convincing proof that is based on the investigation of paths and their de-

11

formation in the complex number plane. This not only proves the theorem, but to a certain degree, explains why it is correct.

We will show, as a consequence of the *Fundamental Theorem of Algebra*, how to factor a polynomial with complex coefficients (including the special case of the real numbers.)

2.1 Paths in the Complex Plane

If point z, in the complex plane P, travels in time t, where $t_0 \leq t \leq t_1$, then we say that we're given a path in the plane P (see Figure 2.1). In this way, the path is a function $z(t)$ of the real variable t; such that $z(t)$, defined on the interval $t_0 \leq t \leq t_1$, can take on the complex values:

$$z(t), \qquad t_0 \leq t \leq t_1$$

Such a formula defines a path. We're talking here about the movement of the point, naturally, without discontinuities, so that the function $z(t)$ is a continuous function. We do not specify the notion of continuity, assuming that it is intuitively clear. It is important to clearly understand that a path is not a curve described by the moving point, rather a path is a process of movement itself. The same curve can be described in more than one way.

In the process of movement, the point $z(t)$, in the different points of time, may end up in the same spot on the plane, therefore, the following equality is allowed:

$$z(t_2) = z(t_3) \quad \text{when} \quad t_2 \neq t_3.$$

In this way, the path can have self-intersections. It may even consist of a single point, specifically in the case, when the point $z(t)$ never moves at all while varying t.

In the following text, unless stated explicitly, we will always assume that the path does not go through the origin, that is the value of $z(t)$ is never zero for any value of t. The point $z(t_0) = z_0$ is called the path's beginning, and the point $z(t_1) =$

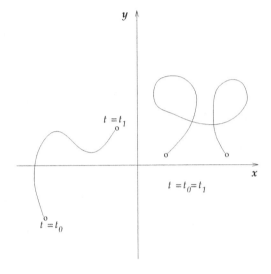

Figure 2.1: Examples of paths.

z_1 the path's end. In case the equality $z_0 = z_1$ holds, then the path is said to be closed (see Figure 2.2).

Since the complex number $z(t)$ is never zero, the argument $\phi(t)$ of the complex number $z(t)$ is defined for any t. But there are infinite number of such arguments, every two of them differ from each other by $2k\pi$. This indefiniteness is undesirable for us. To alleviate it, for the initial point z_0, let us pick a particular argument $\phi_0 = \phi(t_0)$.

As t increases we shall pick the argument $\phi(t)$ of point $z(t)$ in such a way that small changes of t will not change $\phi(t)$ by much. This way the uncertainty in choosing the argument $\phi(t)$ is removed. Addition of $2k\pi$ to the argument, with $k \neq 0$, would immediately result in drastic changes in $\phi(t)$. By choosing the initial value of the argument $\phi(t_0) = \phi_0$, and from then on allowing the argument $\phi(t)$ of the point $z(t)$ to vary continuously with t, we obtain a well defined function $\phi(t)$, that changes continuously, that is, without sudden jumps. Selecting

13

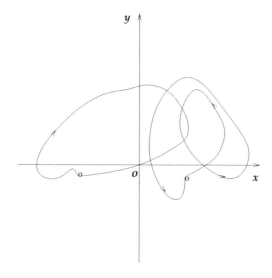

Figure 2.2: Examples of closed paths.

the initial value of the argument ϕ_0 differently, changing it by $2k\pi$, will result in an overall difference of $2k\pi$ for all t. It follows from this that with such an approach to constructing of the $\phi(t)$ function the quantity

$$\phi(t_1) - \phi(t_0) \qquad (2.1)$$

does not depend on the arbitrary choice of the initial value of the argument of the number z_0.

If the path is closed then the points z_0 and z_1 coincide, and therefore, their arguments $\phi(t_0)$ and $\phi(t_1)$ may differ only by $2k\pi$. Thus, the the value of formula (2.1) is exactly $2k\pi$. The whole number k is called the *index* of a closed path in the complex plane. I would like to emphasize that the index of a closed path can only be defined if the path never crosses the zero of the coordinate system.

The index k has a simple geometrical interpretation. It is this index that tells how many times the point $z(t)$, while following the closed path, passes around the origin (see Figure 2.3).

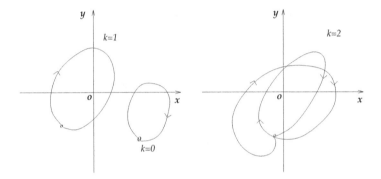

Figure 2.3: Paths with indices 0, 1, 2.

Lets consider a simple example. The path

$$1 + r(cost + isint), \qquad 0 \le t \le 2\pi \tag{2.2}$$

is closed. It describes a circle with center at point 1 and radius r, and the circle is described monotonically in time t in the counterclockwise direction. If the number $r < 1$, then the circle does not contain the origin, and the index of such a path is equal to 0. If $r > 1$ then the interior of the circle contains the zero of the coordinate system, and the index is equal to 1 (check this independently). In case $r = 1$, the path passes through the origin and has an undefined index. If the number r changes then the path (2.2) is said to deform - see Figure 2.4. We can see in this example that during the deformation of the path its index does not change, unless the path, at some point in deformation, passes through the zero of the coordinate system.

15

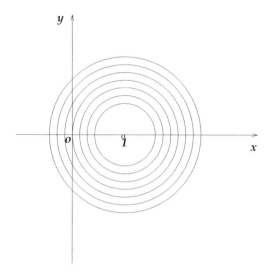

Figure 2.4: Deformation of the path $1+r(\cos\alpha+i\sin\alpha)$ induced by variation of r.

In saying that the path describes the movement of a point, we simply want to give a more intuitive feeling to the definition of a path. In reality, we're talking about the dependence of the complex variable z on the real parameter t (which could be denoted by a different letter). In this way, for example, the path of formula (2.2) can be given as:

$$1 + r(\cos\alpha + i\sin\alpha), \qquad 0 \le \alpha \le 2\pi, \qquad (2.3)$$

where the parameter is now not t, but α. It is clear that the path of formula (2.3) is described by the point $1 + z$, when the point z describes the path $r(\cos\alpha + i\sin\alpha)$, $0 \le \alpha \le 2\pi$. Here r is a numerical parameter that the path depends on. The path of formula (2.3) is said to deform with a change in r.

Let us define the deformation of a path. We will say that the path deforms if it slowly changes without jumps as a function

Figure 2.5: The deformation of a path.

of some parameter, which for the path of Figure 2.3 is r, but in general can be any other letter, for examples, s (see Figure 2.5). In this way, the deforming path is written down as:

$$z(\alpha, s), \qquad \alpha_0 \le \alpha \le \alpha_1, \qquad s_0 \le s \le s_1. \qquad (2.4)$$

Here for every fixed s there exists a specific path that is described when α changes between α_0 and α_1. And with changing s the path changes itself; it deforms. Clearly if the path (2.4) is closed, that is, for every s the equality

$$z(\alpha_0, s) = z(\alpha_1, s)$$

holds, then throughout the deformation, the index should change without jumps. And, since the index is an integer, it remains a constant. Certainly, this is only true in the case when for an arbitrary s the path (2.4) does not pass through the origin. Otherwise for such s the index of the path is not defined. Therefore, we can make the following statement:

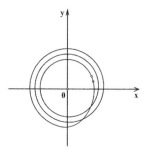

Figure 2.6:

If a closed path continuously deforms, never passing through the origin during the deformation, then its index does not change. Consider another example of a closed path.

$$r^n(\cos(n\alpha) + i\sin(n\alpha)), \qquad 0 \le \alpha \le 2\pi. \qquad (2.5)$$

Clearly this path is described by a point z^n, when the point z describes the closed path

$$r(\cos\alpha + i\sin\alpha), \qquad 0 \le \alpha \le 2\pi.$$

We can see that as α changes from 0 to 2π, the argument of the point z^n changes from 0 to $2n\pi$. Therefore, the index of the path 2.5 is equal to n (see Figure 2.6 where the case for $n = 3$ is shown schematically).

2.2   The Complex Functions of a Complex Variable

If the numerical value of a complex variable w is found by knowing the numerical value of a different complex variable z, then the first variable w is said to be a function of the variable z, which is noted in this way:

$$w = f(z).$$

If the complex function $f(z)$ of a complex argument is differentiable, then such function is said to be *analytical*. Today, the theory of analytical functions is one of the most important branches of mathematics. Here we will be concerned with analytical functions of a very specific form: polynomials.

Consider the following polynomials:

$$w = f(z) = z^n + a_1 z^{n-1} + ... + a_{n-1} z + a_n, \qquad (2.6)$$

where the coefficients a_1, a_2, ..., a_n are complex, and n is a non-negative integer that is called the degree of the polynomial. The goal of our investigation is a proof that the polynomial of equation (2.6) of positive degree has a root. That is, the equation

$$f(z) = 0,$$

where $f(z)$ is a polynomial like equation (2.6) with $n > 0$, has a solution. To prove this, let us consider the closed path

$$f[r(\cos \alpha + i \sin \alpha)], \qquad 0 \le \alpha \le 2\pi, \qquad (2.7)$$

which is described by a point $f(z)$ while the point z describes the path given by

$$r(\cos \alpha + i \sin \alpha), \qquad 0 \le \alpha \le 2\pi.$$

The case when the free parameter a_n of the polynomial $f(z)$ is equal to 0 does not require consideration, as, in such a case, the polynomial $f(z)$ has an obvious root $z = 0$. For this reason, we shall assume that $a_n \ne 0$. The closed path in formula (2.7) depends on the parameter r, and with change in r the path deforms. When $r = 0$ the number z is equal to 0, and the path $f(z)$ consists of a single stationary point a_n. In this way the path's index is 0 when $r = 0$. We shall prove that for r sufficiently large the path's index is equal to n. But, by

19

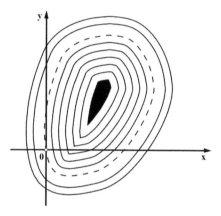

Figure 2.7:

definition $n \neq 0$, thus with the change in r from large to 0 the path of formula (2.7) will deform passing through the origin, therefore, this polynomial has a root (see Figure 2.7).

To prove that with sufficiently large r the path of formula (2.7) has an index equal to n, let's deform this path into a simpler one, where it will be simple to calculate the index.

To begin with, we shall break up the polynomial $f(z)$ into the sum of two polynomials:

$$f(z) = z^n + g(z),$$

where $g(z)$ is given by the formula:

$$g(z) = a_1 z^{n-1} + a_2 z^{n-2} + ... + a_{n-1} z + a_n.$$

Since the coefficients a_1, a_2, ..., a_n of the polynomial $g(z)$ are determinate numbers, the absolute value of any of them is smaller that some constant c. From the proven inequality in 1.7, extended to the sum of a sequence of an arbitrary length, it follows that when $|z| > 1$

$$|g(z)| \leq nc|z^{n-1}|. \tag{2.8}$$

Let us consider the polynomial $f(z, s)$ that depends on the parameter s, $0 \leq s \leq 1$, and is given by the following formula:

$$f(z, s) = z^n + s \cdot g(z).$$

We get the equality

$$z^n = f(z, s) - s \cdot g(z),$$

and from this

$$\begin{aligned} |z^n| &\leq |f(z, s)| + |-s \cdot g(z)| \\ &\leq |f(z, s)| + nc|z^{n-1}|s \\ &\leq |f(z, s)| + nc|z^{n-1}| \end{aligned}$$

where we used the result from formula (2.8). From this it follows that

$$|f(z, s)| \geq |z^n| - nc|z^{n-1}|.$$

If we replace $|z|$ with r, then the last inequality becomes:

$$|f(z, s)| \geq r^n - nc \cdot r^{n-1} = r^{n-1}(r - cn).$$

This way when $r > cn$ the right side of the previous inequality is positive. Therefore, only if $r > cn$ does the absolute value of $f(z, s)$ not become 0 for any value of s.

Now, let us make a note of the fact that when $s = 0$ the polynomial $f(z, s)$ turns into the polynomial we know z^n. And the index of the path z^n, when z describes the circle $r(\cos \alpha + i \cdot \sin \alpha)$, was already calculated earlier in equation (2.5). It is equal to n. When $s = 1$ the polynomial $f(z, s)$ turns into the polynomial $f(z)$, and the path (2.7) defined by it also has an index that is equal to n. Thus, we have proven that the index of the path (2.7), defined by the polynomial $f(z)$ when $r > cn$, is equal to n. In this way, if we consider r varying between 0 and $cn + \epsilon$, where $\epsilon > 0$, the path (2.7) deforms accordingly. Specifically, when $r = 0$, this path becomes a single point and its index becomes equal to 0. And when $r = cn + \epsilon$ its index is equal to n. From this it is evident that in the process of changing r, the path (2.7) for some r goes through the origin, and therefore, the polynomial $f(z)$, for some value of z, $|z| \leq cn + \epsilon$, becomes 0.

Thus, the fundamental theorem of algebra is proven. ∎

3.1 Polynomial Division

Dividing a positive integer a by a positive integer b we arrive at the following equality:

$$a = bh + k, \qquad (3.1)$$

where h and k are non-negative integers, and $k < b$. The number h is called the *quotient*, and k is called the *remainder*.

The method for integer division is well known from arithmetic, but just as with numbers, it is possible to divide polynomials by each other. We begin with two polynomials:

$$a(z) = a_0 z^p + a_1 z^{p-1} + \dots + a_p,$$

$$b(z) = b_0 z^q + b_1 z^{q-1} + \dots + b_q.$$

Let us assume here that a_0 and b_0 are non-zero, therefore, the polynomial $a(z)$ is of degree p, and $b(z)$ is of degree q. Dividing the polynomial $a(z)$ by the polynomial $b(z)$ we will arrive at the following equality, analogous to (3.1):

$$a(z) = b(z)h(z) + k(z), \qquad (3.2)$$

where the degree of $k(z)$ is less than q. The polynomials $h(z)$ and $k(z)$ are called the quotient, and the remainder respectively.

If $k(z) \equiv 0$, then it is said that the polynomial $a(z)$ is exactly divisible by $b(z)$, and $h(z)$ is the division factor.

The equality (3.1) is proven in arithmetic, and the equality (3.2) ought to be proven in algebra. Polynomial division is not part of today's high school program. To prove equality (3.2), we must construct polynomials $h(z)$ and $k(z)$ that satisfy this equality. This construction process is in itself a very important algorithm, the so-called *Euclid's Algorithm*. Let us describe it.

If $p < q$, then $h(z) = 0$, $k(z) = a(z)$ and the equality (3.2) is satisfied.

We will construct the polynomials $h(z)$ and $k(z)$ assuming that $p \geq q$. First we construct the equality

$$a(z) = b(z)h_1(z) + a_1(z), \qquad (3.3)$$

where the degree of the polynomial $a_1(z)$ is less than p. For this we shall assume

$$h_1(z) = \frac{a_0}{b_0}z^{p-q}.$$

Then the difference

$$a(z) - b(z)h_1(z) = a_1(z)$$

has degree less than p, for in this polynomial the coefficient of z^p is zero. All the other powers of z that appear in this polynomial are obviously lower than p. In this way, the equality (3.3) is constructed.

If the polynomial $a_1(z)$ has degree less than q, then the equality (3.3) is already the equality in (3.2). Otherwise, let us apply the same procedure to the polynomial $a_1(z)$ that we have already applied to the polynomial $a(z)$ in the construction of the equality (3.3). Then we obtain the following equality:

$$a_1(z) = b(z)h_2(z) + a_2(z),$$

Notice that the degree of the polynomial $a_2(z)$ is less then the degree of the polynomial $a_1(z)$. If the polynomial $a_2(z)$

has degree less than q, then substituting $a_1(z)$ from the last equality into equality (3.3), we obtain

$$a(z) = b(z)[h_1(z) + h_2(z)] + a_2(z),$$

which now has the same form as the equality (3.2). If the degree of the polynomial $a_2(z)$ is also greater than q, then we continue our construction further. In the end, we will prove the needed equality (3.2).

We described here the process of division of the polynomial $a(z)$ by the polynomial $b(z)$, that is, the way to find polynomials $h(z)$ and $k(z)$ from the equality (3.2). Let us prove now, that the polynomials $h(z)$ and $k(z)$ are uniquely determined by the polynomials $a(z)$ and $b(z)$. Assume that, along with the equality (3.2), another equality is also true:

$$a(z) = b(z)h_0(z) + k_0(z),$$

and the degree of the polynomial $k_0(z)$ is less than q. Subtracting this equality from the equality 3.2, we obtain

$$b(z)[h(z) - h_0(z)] = k_0(z) - k(z).$$

Since the degree of the polynomial $b(z)$ is equal to q, and the degree of the polynomial $k_0(z) - k(z)$ is less that q, then the last equality is true only if $h(z) - h_0(z) = 0$, hence $k(z) - k_0(z) = 0$.

Notice that complex numbers will not be produced during polynomial division. In the case where polynomials $a(z)$ and $b(z)$ have real coefficients, then the polynomials $h(z)$ and $k(z)$ must also have real coefficients.

3.2 Polynomial Decomposition Into Factors

Using the argument of the fundamental theorem of algebra, the existing root of the polynomial

$$f_0(z) = z^n + a_1 \cdot z^{n-1} + \ldots + a_n, \; for \; n > 0$$

can be denoted by α_1. Let us prove that the polynomial $f_0(z)$ is divisible by $(z - \alpha_1)$. Dividing, according to the rules of section 3.1, the polynomial $f_0(z)$ by the first degree polynomial $(z - \alpha_1)$, we obtain a quotient denoted by $f_1(z)$. The division remainder will be a zero degree polynomial, that is a number, which can be denoted by k. Therefore, we have:

$$f_0(z) = f_1(z) \cdot (z - \alpha_1) + k.$$

Since $f_0(\alpha_1) = 0$, then with $z = \alpha_1$ above, the $f_1(\alpha_1)$ term equals 0 so we get $k = 0$.

So, the polynomial $f_0(z)$ is divisible by $(z - \alpha_1)$ and we get

$$f_0(z) = f_1(z) \cdot (z - \alpha_1),$$

where $f_1(z)$ is a $n - 1$ degree polynomial, which, naturally, begins with the term z^{n-1}. If $n > 1$, then $n - 1 > 0$, and the polynomial $f_1(z)$ is of positive degree and has, according to the method proved above, a root α_2. As we have shown, the polynomial $f_1(z)$ can be decomposed into factors

$$f_1(z) = f_2(z) \cdot (z - \alpha_2). \tag{3.4}$$

Continuing with this process we obtain a decomposition of the polynomial $f_0(z)$ into n linear multipliers

$$f_0(z) = (z - \alpha_1)(z - \alpha_2)...(z - \alpha_n).$$

The numbers $\alpha_1, \alpha_2, ...\alpha_n$ are the roots of the polynomial $f_0(z)$, and it is clear that there are no other roots of this polynomial; however, some of the roots in this decomposition may be identical. If we group the identical roots together we get the following formula

$$f_0(z) = (z - \alpha_1)^{k_1}(z - \alpha_2)^{k_2}...(z - \alpha_q)^{k_q}, \tag{3.5}$$

where all the roots are different. The number k_1 is called the multiplicity of the root α_1, the number k_2, the multiplicity of the root α_2, and so on, and the number k_q is the multiplicity of the root α_q. Thus the number of different roots of polynomial $f_0(z)$ could be less than n. However, if we take into account the multiplicity of each root, the sum of these multiplicities will be exactly equal to n. In this sense the polynomial $f_0(z)$ has exactly n roots.

Let us now consider the case when all coefficients of the polynomial $f_0(z)$ are real numbers. We can make some interesting auxiliary statements about roots of such a polynomial.

To investigate this case let us introduce the notion of the number $\bar{z}$ conjugated to a given number z.

In particular, if

$$z = x + iy,$$

then by definition the number

$$\bar{z} = x - iy \tag{3.6}$$

is called the conjugate of z.

Therefore, the number $\bar{z}$ conjugated with a number z is the symmetrical reflection of the number z in the "mirror" of the real number axis.

If

$$z = r(\cos\alpha + i\sin\alpha),$$

then

$$\begin{aligned} \bar{z} &= r(\cos\alpha - i\sin\alpha) = \\ &= r[\cos -\alpha + i\sin -\alpha]. \end{aligned} \tag{3.7}$$

Therefore, arguments of two conjugated numbers differ only in sign. From formula (3.6) it follows that

$$\overline{z_1 + z_2} = \bar{z}_1 + \bar{z}_2$$

(See Chapter 1 (1.2).)

From formula (3.7) it follows that

27

$$\overline{z_1 z_2} = \bar{z}_1 \bar{z}_2.$$

(See Chapter 1 (1.12).)

Let us also notice that equality

$$\bar{z} = z$$

happens only if z is a real number.

From the above three formulas it is easy to deduce that the following equality holds true for a polynomial $f_0(z)$ with real coefficients:

$$\overline{f_0(z)} = f_0(\bar{z}).$$

From this equality it follows that if α_1 is a root of polynomial $f_0(z)$ with real coefficients, then $\bar{\alpha}_1$ is also root.

In case α_1 is a real number, this statement is a tautology. In case α_1 is not a real number, this statement indicates that along with the root α_1 there exists another root $\bar{\alpha}_1$ that is different from α_1. Thus, if α_1 is not a real number, then the polynomial $f_1(z)$ (see (3.4)) is divisible by $(z - \bar{\alpha}_1)$, and we get the following decomposition

$$f_0(z) = (z - \alpha_1)(z - \bar{\alpha}_1)f_2(z).$$

Therefore, we have found a quadratic divisor of the polynomial $f_0(z)$

$$g_2(z) = (z - \alpha_1)(z - \bar{\alpha}_1) =$$
$$= z^2 - (\alpha_1 + \bar{\alpha}_1)z + \alpha_1 \bar{\alpha}_1$$

It is clear that the polynomial $g_2(z)$ has real coefficients, and does not have any real divisors because it does not have any real roots. From that it follows that the polynomial $f_2(z)$ also has real coefficients, because it is the result of division of the polynomial $f_0(z)$ by a quadratic with real coefficients (see the end of section 3.1.) Therefore, if α_1 is a real root, we have the following decomposition

$$f_0(z) = g_1(z)f_1(z).$$

where $g_1(z) = z - \alpha_1$, and if α_1 is not real, then we have the following decomposition

$$f_0(z) = g_2(z)f_2(z).$$

Therefore, in any case $f_0(z)$ is divisible either by a polynomial of the first or the second degree that cannot be divided by any polynomial with real roots.

A polynomial with real coefficients that cannot be divided by a polynomial with real roots and of lower degree is called irreducible.

Thus, polynomial $f_0(z)$ may be decomposed by irreducible polynomials of the first or the second degree.

From this follows an important corollary:

Each root of polynomial $f_0(z)$ is a root of some irreducible polynomial of the first or the second degree.

3.3 Greatest common divisor of two polynomials

Polynomial $b(z)$ is a divisor of polynomial $a(z)$ if it divides $a(z)$ exactly, in other words in formula 3.2 $k(z) \equiv 0$.

Polynomial $b(z)$ is a common divisor of two polynomials $a_0(z)$ and $a_1(z)$, if it is a divisor of both these polynomials.

A common divisor $c(z)$ of two polynomials $a_0(z)$ and $a_1(z)$ is called the greatest common divisor of these two polynomials, if it can be divided exactly by any common divisor $b(z)$ of the polynomials $a_0(z)$, and $a_1(z)$.

Existence of the greatest common divisor of two polynomials is not obvious. The existence of the greatest common divisor will be proved below by means of successive division of polynomials, i.e. using operations that are easy to perform, if we do not take into account the complexity of the calculations. But first of all let us prove that two greatest common divisors $c_0(z)$ and $c_1(z)$ of two polynomials $a_0(z)$ and $a_1(z)$ are, basically, the same. In essence, they can differ only by a numeral factor that is not equal to zero. Let us prove it.

Since $c_0(z)$ is the greatest common divisor of polynomials $a_0(z)$ and $a_1(z)$, and $c_1(z)$ is a common divisor of these polynomials, $c_0(z)$ can be divided by $c_1(z)$ exactly, thus we have the following identity

$$c_0(z) = h_1(z)c_1(z). \tag{3.8}$$

By analogy we have the following formula

$$c_1(z) = h_0(z)c_0(z). \tag{3.9}$$

Substituting $c_1(z)$ from formula (3.8) by its expression from formula (3.9), we get

$$c_0(z) = h_0(z)h_1(z)c_0(z).$$

When two polynomials are multiplied, their degrees are added, therefore, from the last formula it follows that polynomials $h_0(z)$ and $h_1(z)$ have degrees equal to zero, that is, they are numbers.

Thus it is proven that there is only one greatest common divisor of two polynomials $a_0(z)$ and $a_1(z)$.

Let us move on to construction of the greatest common divisor $c(z)$ of polynomials $a_0(z)$ and $a_1(z)$. To construct it let us divide polynomial $a_0(z)$ by polynomial $a_1(z)$. Let us write formula (3.2) in the following way

$$a_0(z) = a_1(z)h_1(z) + a_2(z), \tag{3.10}$$

i.e. let us identify the reminder as $a_2(z)$. Now, let us divide polynomial $a_1(z)$ by $a_2(z)$, and write formula (3.2) in the form

$$a_1(z) = a_2(z)h_2(z) + a_3(z), \tag{3.11}$$

in other words, identify the reminder as $a_3(z)$. Now, let us divide polynomial $a_2(z)$ by $a_3(z)$, and write formula (3.2) in the form

$$a_2(z) = a_3(z)h_3(z) + a_4(z),$$

Since the degree of the reminder in this process is constantly decreasing, at some point either the degree of the reminder, or the reminder itself will become zero. If a polynomial is divided by a polynomial which has degree zero, the reminder, of course, will be zero. Thus, eventually we get the following identities

$$a_{n-2}(z) = a_{n-1}(z)h_{n-1}(z) + a_n(z), \qquad (3.12)$$

$$a_{n-1}(z) = a_n(z)h_n(z) \qquad (3.13)$$

If now $b(z)$ is a common divisor of polynomials $a_0(z)$ and $a_1(z)$, then from formula (3.10) it follows that it is a divisor of $a_2(z)$. From formula (3.11) follows that polynomial $b(z)$ is a divisor of $a_3(z)$. In this way we may prove that polynomial $b(z)$ divides exactly all polynomials we have constructed

$$a_0(z), a_1(z), a_2(z), ..., a_n(z),$$

and in particular, it is a divisor of the polynomial $a_n(z)$. From formula (3.13) it follows that the polynomial $a_n(z)$ is a divisor of $a_{n-1}(z)$, from formula (3.12) it follows that it is a divisor of $a_{n-2}(z)$. In the end we prove that $a_n(z)$ is a divisor of polynomials $a_0(z)$ and $a_1(z)$. Thus, we proved that $a_n(z)$ is divided exactly by any divisor of $a_0(z)$ and $a_1(z)$, and is itself a divisor of these polynomials. Therefore, polynomial $a_n(z)$ is the greatest common divisor of the polynomials $a_0(z)$ and $a_1(z)$.

Let us prove now the following important result, which has numerous applications in algebra.

The greatest common divisor $c(z)$ of two polynomials $a_0(z)$ and $a_1(z)$ may be written as follows

$$c(z) = p_0(z)a_0(z) + p_1(z)a_1(z), \qquad (3.14)$$

where $p_0(z)$, $p_1(z)$ are some polynomials. Let us prove this statement. Replacing polynomial $a_2(z)$ in formula (3.11) by its expression from formula (3.10), we get

$$a_3(z) = p_3(z)a_0(z) + q_0(z)a_1(z).$$

If we carry on this process, we will get formula (3.14). Therefore, formula (3.14) has been proven. ∎

3.4 Eliminating multiple roots

Finding roots of a polynomial is one of the oldest and most difficult problems of algebra. Initial attempts to solve it used the formula composed of algebraic expressions that included extraction of roots. It was most adeptly done for the degree two polynomials. For the degree three polynomials there also is a formula of that kind, but it has much less practical significance. There also is a formula for the degree four polynomials, but it completely lacks any practical significance. Later it was proved that for polynomials of degree higher than four a formula for finding roots in radicals does not exist. The complete theory of the possibility of finding roots of polynomials using radicals was created by Galois. On the path to solving the problem of finding roots of polynomials lies the much simpler problem: for a polynomial $f(z)$ that has both simple and multiple roots, find the polynomial $g(z)$ that has the same roots, but all of them are simple. This problem can be solved using Euclid's algorithm, in other words by dividing polynomials by each other. A solution of this problem will be provided in this section.

Let $a(z)$ and $b(z)$ be two polynomials, and let $p(z)$ be their product,

$$p(z) = a(z)b(z).$$

Let us denote as

$$\alpha_1, \alpha_2, ..., \alpha_q$$

the set of numbers, each of which is a root of either $a(z)$, or $b(z)$. If so, these polynomials can be written in the following way:

$$a(z) = (z - \alpha_1)^{k_1}(z - \alpha_2)^{k_2}...(z - \alpha_q)^{k_q}, \qquad (3.15)$$

$$b(z) = (z - \alpha_1)^{l_1}(z - \alpha_2)^{l_2}...(z - \alpha_q)^{l_q}. \qquad (3.16)$$

In each of these decompositions some of the powers may be zero. If so we will consider that the multiplicity of the root is equal to zero. Clearly, we have

$$p(z) = (z - \alpha_1)^{k_1+l_1}(z - \alpha_2)^{k_2+l_2}...(z - \alpha_q)^{k_q+l_q}$$

Thus the multiplicities of each root are added when the polynomials are multiplied. Using the decomposition of polynomials (3.15) and (3.16), we can easily find their greatest common divisor $c(z)$. Let us denote m_1 as the minimum of k_1 and l_1, m_2 as the minimum of k_2 and l_2, and so on, m_q as the minimum of k_q and l_q. Then the greatest common divisor is

$$c(z) = (z - \alpha_1)^{m_1}(z - \alpha_2)^{m_2}...(z - \alpha_q)^{m_q}.$$

This simple way of finding the greatest common divisor of two polynomials requires, however, finding the roots of both polynomials, and therefore, from the practical standpoint, is almost useless, because it turns a simple problem into a very difficult one. Meanwhile, as we saw, finding the greatest common divisor of two polynomials is a simple problem that can be solved using Euclid's algorithm, i. e. by dividing polynomials by each other.

Let us find out now how the multiplicities of the roots of polynomials $a_0(z)$ and $a_1(z) = a_0'(z)$ are interconnected, in other words how the multiplicities of the roots of polynomial $a_0(z)$ and its first derivative $a_0'(z)$ are interconnected.

It turns out that if a root α of polynomial $a_0(z)$ has multiplicity equal to k, then the same root of the $a_0(z)$ first derivative has multiplicity equal to $k-1$. Of course besides the roots that are common to both $a_0(z)$ and $a_0'(z)$, the $a_0'(z)$ polynomial may have some other roots.

Let us prove that if α is a root of polynomial $a_0(z)$ with multiplicity equal to k, the polynomial $a_1(z) = a_0'(z)$ has the same root, but with multiplicity equal to $k-1$.

From the decomposition (3.5) of polynomial $a_0(z)$, it follows that if α is a root of polynomial $a_0(z)$, and its multiplicity is equal to k, then

$$a_0(z) = (z - \alpha)^k b(z),$$

note that $b(z)$ cannot be divided exactly by $(z - \alpha)$, and therefore, α is not its root. Let us differentiate the last equation, we get

$$a_0'(z) = k(z - \alpha)^{k-1} b(z) + (z - \alpha)^k b'(z), \qquad (3.17)$$

$$(z - \alpha)^{k-1} b(z) = \frac{1}{k} a_1(z) - \frac{1}{k}(z - \alpha)^k b'(z) \qquad (3.18)$$

It is clear from formula (3.17) that $(z - \alpha)^{k-1}$ is a divisor of $a_1(z)$. On the other hand, it is also clear from formula (3.18) that $(z - \alpha)^k$ is not a divisor of $a_1(z)$. Indeed, if $a_1(z)$ can be divided exactly by $(z - \alpha)^k$, then it is also possible to divide $b(z)$ by $(z - \alpha)$ exactly. Therefore, the multiplicity of root α of polynomial $a_0'(z)$ is $k-1$. In particular, if α is a single root of polynomial $a_0(z)$, it is not a root of polynomial $a_0'(z)$.

Therefore, if $a_0(z)$ has the decomposition (3.5) then the greatest common divisor of polynomials $a_0(z)$ and $a_1'(z)$ is the polynomial

$$c(z) = (z - \alpha_1)^{k_1-1}(z - \alpha_2)^{k_2-1}...(z - \alpha_q)^{k_q-1}.$$

From this formula it is clear that polynomial $a_0(z)$ can be divided exactly by polynomial $c(z)$, and the quotient is

$$\frac{a_0(z)}{c(z)} = (z - \alpha_1)(z - \alpha_2)...(z - \alpha_q)$$

Therefore, the problem that we posed at the beginning of this section about finding the polynomial $g(z)$ that has the same roots as polynomial $f(z)$ with all of the roots having multiplicity one is solved.

3.5 The number of real roots inside a given interval

Finding a polynomial's complex roots is a problem more difficult than finding its real roots, because in the case of a complex root we have to solve an equation with two unknown variables, one of them a real part of the root and another, its imaginary part.

Since we found the method of reducing the problem of calculating polynomial roots to the problem of calculating roots of another polynomial that has the same roots as the first one, but only with roots of multiplicity one, we are going to investigate this particular case when a polynomial has only simple roots. We are going to achieve a result that allows us to avoid a good deal of approximate calculation of a polynomial's real roots.

Let us consider the polynomial $a_0(x)$ of real variable x that has only simple roots, and let us make up a method of finding out how many roots lie in the given interval $\xi_0 \leq x \leq \xi_1$. Since all roots of polynomial $a_0(x)$ are simple ones, polynomials $a_0(x)$ and $a_1(x) = a_0'(x)$ do not have any common roots.

To solve this problem, let us consider the sequence of real numbers

$$a_0, a_1, a_2, ..., a_n. \tag{3.19}$$

We will assume that none of these numbers is zero. The meaning of sign changing in the sequence (3.19) is almost self evident, but let us define it in a formal way. We will assume that between a_0 and a_1 is a sign change if one of these numbers is negative and the other one is positive. If this is the case, let us set p_1 to 1. If both of a_0 and a_1 are either positive or negative, we will assume that there is no sign change between them. In that case let us set p_1 to zero. In the same way let us define number p_2 that describes the sign change between a_1 and a_2. Carrying this process on we get a sequence of ones and zeros

$$p_1, p_2, ..., p_n.$$

The sum of the numbers in this sequence is the number of sign changes in the sequence (3.19).

Using polynomials $a_0(x)$ and $a_1(x) = a_0'(x)$ as a starting point let us construct a sequence of polynomials.

Let us divide the polynomial $a_0(x)$ by the polynomial $a_1(x)$, and write formula (3.2) as

$$a_0(x) = h_1(x)a_1(x) - a_2(x).$$

Let us now divide the polynomial $a_1(x)$ by the polynomial $a_2(x)$ and write formula (3.2) as

$$a_1(x) = h_2(x)a_2(x) - a_3(x).$$

On the kth step of this process we will get the identity

$$a_{k-1}(x) = h_k(x)a_k(x) - a_{k+1}(x). \tag{3.20}$$

Continuing this process until the end, we obtain a sequence of polynomials

$$a_0(x), a_1(x), ..., a_n(x). \tag{3.21}$$

This process is almost identical with the process of finding the greatest common divisor of two polynomials. Since the polynomials $a_0(x)$ and $a_n(x)$ are relatively prime, the last member of the sequence, namely $a_n(x)$ is a number.

Note, first of all, that there is no such $x = x_0$ that produces zero when substituted into any two successive polynomials of the sequence (3.21). Indeed, if $a_k(x_0) = a_{k+1}(x_0)$, then from formula (3.20) it follows that $a_{k-1}(x_0) = 0$.

Carrying on this process, we eventually come to the conclusion that polynomials $a_0(x)$ and $a_1(x)$ become zero at $x = x_0$, but this is impossible, because these polynomials do not have any common roots.

Now, let x increase, starting from the value ξ_0 and ending at ξ_1. It is clear that in the process of increasing x the number of sign changes of sequence (3.21) also may change. That can happen only when x comes through such an x_0 that causes one of the members of the sequence (3.21) to change its sign, in other words, one of the members passing through zero. Suppose that at $x = x_0$ we have $a_k(x_0) = 0$, and while x increases in the close vicinity of the point x_0 the sign of the function $a_k(x)$ is changing. Assuming that in equation (3.20) $x = x_0$, we get the following equation

$$a_{k-1}(x_0) = -a_{k+1}(x_0).$$

Therefore, the numbers $a_{k-1}(x_0)$ and $a_{k+1}(x_0)$ have opposite signs. If at some x close to x_0, number $a_k(x)$ has sign identical to the sign of $a_{k-1}(x)$, then signs of the numbers $a_k(x)$ and $a_{k+1}(x)$ are different. Therefore, there is no sign change between $a_{k-1}(x)$ and $a_k(x)$, but between $a_k(x)$ and $a_{k+1}(x)$ there is a change of sign. If when the point of x is passing through the point x_0 the sign of the number $a_k(x)$ changes, then this can only happen when there is a sign change between $a_{k-1}(x_0)$ and $a_k(x)$, but there is no sign change between $a_k(x)$ and $a_{k+1}(x_0)$. Therefore, we can assert that while the point of x is passing through the point x_0, so that $a_k(x_0) = 0$, the number of sign changes in the following three member sequence

$$a_{k-1}(x), a_k(x), a_{k+1}(x)$$

stays the same. The only important property of the index k here is that there is a previous member, with index $k - 1$, and next member, with index $k + 1$. Therefore, the number of sign changes of the sequence (3.21) may change only while x is passing through x_0, either the very first or the very last member of sequence 3.21 changes its sign. But the last member can not change its sign because it is a constant.

Let us see now how the number of sign changes will change in the sequence (3.21) for the point x_0, such that the first member of the sequence (3.21) turns zero at this point. If for x close to x_0, but less than x_0, $a_0(x) > 0$, then when x goes through the point x_0, function $a_0(x)$ changes its sign. But at $x = x_0$ the number $a_1(x_0)$ will be negative, because at that point function $a_0(x)$ is decreasing, and therefore, its derivative $a_1(x)$ is negative. Therefore, for x close to x_0, but less than x_0, between $a_0(x)$ and $a_1(x)$ there is a sign change, but for x greater than x_0, but close to x_0, between $a_0(x)$ and $a_1(x)$ there is no sign change. Therefore, when the point of x passes through x_0, at which value function $a_0(x)$ turns to zero, and the function is decreasing, the sign change between $a_0(x)$ and $a_1(x)$ vanishes. In the same way we can prove that if function $a_0(x)$ is increasing when x passes through the point x_0, there is also no sign change. Therefore, we have found that the number of sign changes of sequence (3.21), when x is changing its value, may happen only when x is passing through a root of $a_0(x)$, and the number of sign changes decreases by 1 at that point. Thus when x increases from ξ_0 to ξ_1, sequence (3.21) loses one sign change when x passes through the root of $a_0(x)$. And, therefore, to find the number of roots of polynomial $a_0(x)$ between ξ_1 and ξ_1 we need to compare the following sequences of numbers

$$a_0(\xi_0), a_1(\xi_0), ..., a_n(\xi_0), \tag{3.22}$$

and

$$a_0(\xi_1), a_1(\xi_1), ..., a_n(\xi_1), \qquad (3.23)$$

The number of sign changes of sequence (3.22) can be only greater than the number of sign changes of the sequence (3.23).

The difference of sign changes between sequences (3.22) and (3.23) is the same as the number of roots of polynomial $a_0(x)$ that lie in the interval $\xi_0 \le x \le \xi_1$.

CHAPTER 4

I will precede a description of quaternions with some information about vector spaces, which I will need for examination of quaternions because quaternions constitute a four dimensional Euclidean vector space.

4.1 Vector spaces

A sequence

$$\vec{x} = (x^1, x^2, ..., x^n) \tag{4.1}$$

of n real numbers, in the order stated by formula 4.1, is called an n-dimensional vector. A set A^n of all n-dimensional vectors is called an n-dimensional vector space. The numbers

$$x^1, x^2, ..., x^n$$

are called coordinates of the vector $\vec{x}$ (see (4.1)).

A vector is considered to be equal to 0 and is written as 0 if all its coordinates are zeros.

An arbitrary vector $\vec{x}$ from A^n may be multiplied by a real number α. In order to get the product, each coordinate of the vector x should be multiplied by α. Therefore, a formula for $\alpha\vec{x}$ may be written in the following way:

$$\alpha\vec{x} = (\alpha x^1, \alpha x^2, ..., \alpha x^n). \tag{4.2}$$

Two vectors $\vec{x}$ (see (4.1)) and

$$\vec{y} = (y^1, y^2, ..., y^n)$$

from A^n can be added. In order to add them, their corresponding coordinates are added. This can be written in the following way:

$$\vec{x} + \vec{y} = (x^1 + y^1, x^2 + y^2, ..., x^n + y^n). \qquad (4.3)$$

If now

$$\vec{x}_1, \vec{x}_2, ..., \vec{x}_k \qquad (4.4)$$

is a subset of k vectors of A^n, then using operations (4.2) and (4.3) we can construct their linear form

$$\vec{z} = \alpha_1 \vec{x}_1 + \alpha_2 \vec{x}_2 + ... + \alpha_k \vec{x}_k, \qquad (4.5)$$

where $\alpha_1, ..., \alpha_k$ are real numbers

A set of k vectors (4.4) is linearly independent, if the vector, defined by the formula (4.5), is zero only on the condition that

$$\alpha_1 = 0, \alpha_2 = 0, ...\alpha_k = 0.$$

Otherwise, if there are numbers

$$\alpha_1, \alpha_2, ..., \alpha_k, \qquad (4.6)$$

such that not all of them are equal to 0, and the vector z defined by the formula (4.5) is zero, then set (4.4) is linearly dependent. If to a set of linearly dependent vectors (4.4) we add some number of vectors $\vec{x}_{k+1}, \vec{x}_{k+2}, ..., \vec{x}_l$, then the expanded set of vectors

$$\vec{x}_1, \vec{x}_2, ..., \vec{x}_l$$

will also be linearly dependent. Indeed, if a set of coefficients (4.6) defines a linear dependence of vectors (4.4), then we have the following identity

$$\alpha_1 \vec{x}_1 + \alpha_2 \vec{x}_2 + ... + \alpha_k \vec{x}_k + 0\vec{x}_{k+1} + ... + 0\vec{x}_l = 0,$$

and not all the numbers in the sequence

$$\alpha_1, \alpha_2, ..., \alpha_k, \alpha_{k+1} = 0, ..., \alpha_l = 0$$

are equal to zero.

If in the vector space A^n that comprises all vectors of the kind (4.1) we take all vectors whose ith coordinate is equal to zero, we will get a vector space A^{n-1} of dimension $n-1$ which is called a coordinate subspace of the vector space A^n.

Let us prove the following important statement.

A) A condition for linear dependence.

A set of k vectors (4.4) of n-dimensional space A^n is always linearly dependent if $k > n$.

To prove this statement it is sufficient to prove the special case $k = n + 1$. Let us prove it using the method of induction. First of all, note that for $n = 1$ the statement A) is true. Let $\vec{x}$ and $\vec{y}$ be two vectors of the vector space A^1, so that

$$\vec{x} = (x^1), \ \vec{y} = (y^1).$$

If the numbers x^1, y^1 are equal to zero, we have identity

$$\alpha\vec{x} + \beta\vec{y} = 0$$

for any α and β. So in that case vectors $\vec{x}$ and $\vec{y}$ are linearly dependent. If at least one of x^1 and y^1 is not equal to zero, then we have the identity

$$y^1\vec{x} - x^1\vec{y} = 0$$

43

Therefore, for $n = 1$ the statement A) has been proved. ■
Let now

$$\vec{x}_1, \vec{x}_2, ..., \vec{x}_{n+1} \qquad (4.7)$$

be a set of $n+1$ vectors of the vector space A^n. Let us focus on the last coordinate of all the vectors that form set (4.7). If all these coordinates are equal to zero, then set (4.7) belongs to an $n-1$-dimensional vector space. According to the induction assumption, set (4.7) is linearly dependent. Suppose then that at least one of the vectors, let it be $\vec{x}_{n+1}$ has the last coordinate not equal to zero. In that case for each vector $\vec{x}_j$ from the set (4.7) for $j \leq n$ we can find numbers α_j such that all the vectors $\vec{y}_j$ defined by the formula

$$\vec{y}_j = \vec{x}_j - \alpha_j \vec{x}_{n+1}, j = 1, ..., n, \qquad (4.8)$$

have their last coordinate equal to zero, and therefore, belong to a $(n-1)$-dimensional vector space. Thus according to the induction assumption they are linearly dependent. Therefore, there exists a set of numbers

$$\beta_1, \beta_2, ..., \beta_n$$

(not all of them equal to zero) such that there is an identity

$$\beta_1 \vec{y}_1 + \beta_2 \vec{y}_2 + ... + \beta_n \vec{y}_n = 0.$$

From this formula, and from formula (4.8) it follows that

$$\beta_1 \vec{x}_1 + \beta_2 \vec{x}_2 + ... + \beta_n \vec{x}_n - (\beta_1 \alpha_1 + \beta_2 \alpha_2 + ... + \beta_n \alpha_n) \vec{x}_{n+1} = 0.$$

Therefore, vector set (4.7) is linearly dependent. Hence, the statement A) is proved. ■

B) The basis of a vector space.

Let us designate $\vec{e}_i$ as a vector of space A^n, such that all its coordinates are equal to zero except one coordinate, with the ith element equal to 1. In other words, let us label

$$\vec{e}_1 = (1, 0, ..., 0),$$

$$\vec{e}_2 = (0, 1, ..., 0),$$

$$...,$$

$$\vec{e}_n = (0, 0, ..., 1),$$

It is clear that any vector x (see (4.1)) can be written as

$$\vec{x} = x^1 \vec{e}_1 + x^2 \vec{e}_2 + ... + x^n \vec{e}_n$$

the coefficients $x^1, x^2, ..., x^n$ here are uniquely defined by the vector x. Therefore, every vector x in the vector space A^n may be represented by a linear expression made up of the vectors

$$\vec{e}_1, \vec{e}_2, ..., \vec{e}_n, \tag{4.9}$$

and coefficients which are uniquely defined for each vector. A set of vectors that have this property is called a basis of the vector space A^n. The basis we built (4.9) is not the only one. It turns out that any linearly independent set

$$\vec{\varepsilon}_1, \vec{\varepsilon}_2, ..., \vec{\varepsilon}_n, \tag{4.10}$$

of n vectors of vector space A^n is a basis of this vector space.

Indeed, let $\vec{x}$ be a some vector of the vector space A^n. Then the vectors

$$\vec{x}, \vec{\varepsilon}_1, \vec{\varepsilon}_2, ..., \vec{\varepsilon}_n$$

are linearly dependent because there are $n + 1$ of them (see statement A)). Therefore, we have the following equation

$$\alpha \vec{x}_1 + \alpha_1 \vec{\varepsilon}_1 + \alpha_2 \vec{\varepsilon}_2 + ... + \alpha_n \vec{\varepsilon}_n = 0 \tag{4.11}$$

and not all of the coefficients in the expression are equal to zero. In particular, the coefficient α cannot be zero, because otherwise it would mean that the vectors of set (4.10) are linearly dependent. Keeping this in mind, let us divide both parts of equation (4.11) by α. In order to preserve the same notation, let us think that $\alpha = -1$, and keep all other coefficients as is. In this way, we get

$$\vec{x} = \alpha_1 \vec{\varepsilon}_1 + \alpha_2 \vec{\varepsilon}_2 + ... + \alpha_n \vec{\varepsilon}_n. \qquad (4.12)$$

If vector $\vec{x}$ could be written using basis (4.10) but with different coefficients, we would have the following relation

$$\vec{x} = \beta_1 \vec{\varepsilon}_1 + \beta_2 \vec{\varepsilon}_2 + ... + \beta_n \vec{\varepsilon}_n,$$

Subtracting this last relation from (4.12), we get

$$(\alpha_1 - \beta_1)\vec{\varepsilon}_1 + (\alpha_2 - \beta_2)\vec{\varepsilon}_2 + ... + (\alpha_n - \beta_n)\vec{\varepsilon}_n = 0,$$

That can only be true, in view of linear independence of the vector set (4.10), if $\alpha_i = \beta_i$ for $i = 1, 2, ..., n$.

C) The extension of a linear set of vectors into a basis for A^n.

Let

$$\vec{x}_1, \vec{x}_2, ..., \vec{x}_p, \qquad (4.13)$$

be some linearly independent set of vectors of A^n. If $p < n$, then set (4.13) can be extended by some set of vectors $\vec{x}_{p+1}, ..., \vec{x}_n$ so that the set

$$\vec{x}_1, \vec{x}_2, ..., \vec{x}_p, \vec{x}_{p+1}, ..., \vec{x}_n$$

is linearly independent.

Let us prove the statement C). If C) is not true, then, expanding set (4.13) by vector $\vec{e}_1$ (see B)) we will get a linearly dependent, extended set

$$\vec{e}_1, \vec{x}_1, ..., \vec{x}_p.$$

The vector $\vec{e}_1$ is not zero. Therefore, there is a vector $\vec{x}_q$, that can be written as a linear expression of the vectors of the expanded set. Let us remove it. Then any vector of A^n can be presented as a linear expression of the remaining vectors of the expanded set. Now, let us add to the remaining vectors of the expanded set vector $\vec{e}_2$, we get a linearly dependent set from which we can delete one of the vectors $\vec{x}_1, ..., \vec{x}_{q-1}, \vec{x}_{q+1}, ..., \vec{x}_p$. Carrying on this process we will exclude from the expanded set all vectors (4.13), which means that any vector of A^n can be represented as a linear combination of vectors $\vec{e}_1, ..., \vec{e}_p$, which is not true. Therefore, statement C) is true.

One can see from statement B) in the beginning of the section that a given definition of the n-dimensional vector space A^n is not invariant, because it depends on an arbitrarily chosen basis (4.9), and there are an infinite number of bases in A^n. Let us give another, but invariant, definition of the n-dimensional vector space A^n.

D) A vector space is a set A of elements, with two defined operations, addition and multiplication by real numbers, that satisfy the following condition: if α and β are two real numbers, and $\vec{x}$ and $\vec{y}$ are two elements of A (in other words, two vectors), then the following relations hold

$$(\alpha + \beta)\vec{x} = \alpha\vec{x} + \beta\vec{x},$$

$$\alpha(\vec{x} + \vec{y}) = \alpha\vec{x} + \alpha\vec{y}.$$

Using these operations it is possible to make a linear combination of an arbitrary set of vectors

$$\vec{x}_1, ..., \vec{x}_n,$$

in other words, to make a vector

$$\vec{z} = \alpha_1\vec{x}_1 + \alpha_2\vec{x}_2 + ... + \alpha_n\vec{x}_n,$$

where $\alpha_1, ..., \alpha_n$ are real numbers .

Therefore, it is possible to introduce the notion of linear dependence and independence, in the same way as it was done before (see statement B)).

If in a vector space A there are n linear independent vectors, but no $n + 1$ linear independent vectors, then we will say that the vector space A is of n dimensions, and is named as A^n.

Choosing n linearly independent vectors from the vector space A^n, we get a basis, and using this basis we will get a coordinate notation of any vector of A^n.

E) A vector subspace. Let A be some vector space, and B be a subset of A such that for any vectors $\vec{x}$ and $\vec{y}$ from B, the vector $\vec{x} + \vec{y}$ also belongs to B, and for any vector $\vec{x}$ from B and any real number α the vector $\alpha\vec{x}$ also belongs to B. If all this is true, then clearly B is a vector space and a subspace of A in terms of the operations of A. The set B is called a vector subspace of A.

If a vector space A is of n dimensions, then any subspace of A has dimension less than or equal to n.

F) Let A^n be an n-dimensional vector space and B^p and C^q be two subspaces of dimensions p and q correspondingly. If the only common vector of B^p and C^q is zero, and $p + q = n$, then the vector space A^n partitions into the Cartesian product of its vector subspaces B^p and C^q. Specifically, each vector $\vec{x}$ of A^n may be represented as a sum

$$\vec{x} = \vec{y} + \vec{z},$$

where $\vec{y}$ is a vector from B^p, and $\vec{z}$ is a vector from C^q, and for any $\vec{x}$ there is only one such $\vec{y}$, and only one such $\vec{z}$.

To prove this statement let us choose a basis of subspace B^p

$$\vec{\varepsilon}_1, \vec{\varepsilon}_2, ..., \vec{\varepsilon}_p \tag{4.14}$$

that comprises p elements, and a basis of subspace C^q

$$\vec{\varepsilon}_{p+1}, \vec{\varepsilon}_{p+2}, ..., \vec{\varepsilon}_n \tag{4.15}$$

that comprises q elements. Combining sets (4.14) and (4.15), we get a set

$$\vec{\varepsilon}_1, \vec{\varepsilon}_2, ..., \vec{\varepsilon}_n \qquad (4.16)$$

Let us prove that this set is a basis of A^n. Because the number of vectors in set (4.16) is equal to the dimension of A^n, it is enough just to prove that the vectors of the set (4.16) are linearly independent. Let us suppose that it is not true. In particular, then the following identity is true

$$\alpha_1\vec{\varepsilon}_1 + \alpha_2\vec{\varepsilon}_2 + \alpha_p\vec{\varepsilon}_p + \alpha_{p+1}\vec{\varepsilon}_{p+1} + ... + \alpha_n\vec{\varepsilon}_n = 0 \qquad (4.17)$$

and not all of the coefficients α_i are equal to zero. Let us designate $\vec{y}$ as the sum of the first p elements of (4.17), and $\vec{z}$ as the sum of the remaining q elements. We get two vectors $\vec{y}$ and $\vec{z}$, both of which cannot be zero, so at least one of them is not zero. We may re-write (4.17) as

$$\vec{y} + \vec{z} = 0,$$

or as

$$\vec{y} = -\vec{z},$$

where $\vec{y} \in B^p$, and $\vec{z} \in C^q$.

Therefore, we come to the conclusion that vector subspaces B^p and C^q have a common vector not equal to zero. This is a contradiction.

Therefore, set 4.16 is a basis of vector space A^n, and every vector of this space can be written as

$$\vec{x} = \alpha_1\vec{\varepsilon}_1 + \alpha_2\vec{\varepsilon}_2 + \alpha_p\vec{\varepsilon}_p + \alpha_{p+1}\vec{\varepsilon}_{p+1} + ... + \alpha_n\vec{\varepsilon}_n$$

If we name as $\vec{y}$ the sum of the first p elements of this formula, and $\vec{z}$ as the sum of the remaining q elements, we get

$$\vec{x} = \vec{y} + \vec{z}$$

where $\vec{y} \in B^p$, and $\vec{z} \in C^q$. Therefore, the statement F) has been proved. ∎

G) Let X and Y be two vector spaces, and let f be a mapping that maps each vector $\vec{x} \in X$ to some vector $f(\vec{x}) \in Y$, in a such way that

$$f(\vec{x}_1 + \vec{x}_2) = f(\vec{x}_1) + f(\vec{x}_2),$$

$$f(\alpha\vec{x}) = \alpha f(\vec{x}),$$

where α is any real number. A mapping that has such properties is called a linear mapping.

We will say that it is a mapping of space X *onto* space Y, instead of saying that it is a mapping of space X *into* space Y, if for every vector in space Y there is at least one vector in space X that maps into the given vector of space Y.

A mapping f is isomorphic, if the zero of space X maps to the zero of space Y; in other words, if from relation $f(\vec{x}) = 0$ it follows that $\vec{x} = 0$. In that case, f is a one to one mapping of X onto the subspace $f(X)$ of the space Y. Indeed, two nonidentical vectors $\vec{x}_1$ and $\vec{x}_2$ of the space X cannot be mapped on the same vector of the space Y, because in that case the vector $\vec{x}_1 - \vec{x}_2 \neq 0$ would be mapped on the zero of the space Y.

4.2 Euclidean vector spaces

A vector space A^n is called a Euclidean vector space if the scalar product operation is defined in that space.

A) A scalar product operation is defined in vector space A^n if for any pair of vectors $\vec{x}$ and $\vec{y}$ of A^n there is a corresponding real number labeled as $(\vec{x}, \vec{y})$, and called the scalar product of vectors $\vec{x}$ and $\vec{y}$. This operation complies with the conditions of symmetry and linearity. The symmetry condition means that the following relationship must occur

$$(\vec{x}, \vec{y}) = (\vec{y}, \vec{x}) \tag{4.18}$$

The linearity condition means that if $\vec{x}_1$ and $\vec{x}_2$ are two vectors of A^n, and α is a real number, then the following relations are true

$$(\vec{x}_1 + \vec{x}_2, \vec{y}) = (\vec{x}_1, \vec{y}) + (\vec{x}_2, \vec{y}),$$

$$(\alpha\vec{x}, \vec{y}) = \alpha(\vec{x}, \vec{y})$$

Because of symmetry, the linearity condition is also true for the second vector $\vec{y}$. In other words:

$$(\vec{x}, \vec{y}_1 + \vec{y}_2) = (\vec{x}, \vec{y}_1) + (\vec{x}, \vec{y}_2),$$

$$(\vec{x}, \beta\vec{y}) = \beta(\vec{x}, \vec{y}).$$

Let

$$\vec{\varepsilon}_1, \vec{\varepsilon}_2, ..., \vec{\varepsilon}_n$$

be some basis of space A^n, therefore,

$$x = x^1\vec{\varepsilon}_1 + x^2\vec{\varepsilon}_2, +... + x^n\vec{\varepsilon}_n$$

$$y = y^1\vec{\varepsilon}_1 + y^2\vec{\varepsilon}_2, +... + y^n\vec{\varepsilon}_n$$

Let us designate

$$(\vec{\varepsilon}_i, \vec{\varepsilon}_j) = g_{i,j},$$

Because of symmetry, we have

$$g_{i,j} = g_{j,i},$$

Because of linearity, we have

$$(\vec{x}, \vec{y}) = \sum_{i,j=1}^{n} x^i y^j g_{i,j} = \sum_{i,j=1}^{n} g_{i,j} x^i y^j \qquad (4.19)$$

This is the coordinate notation for a scalar product. There is one more very important property of a scalar that is included in the definition.

The scalar power of a vector, in other words the scalar product of two identical vectors, is considered a power of two of the vector lengths, and therefore, is always greater or equal to zero, and is zero only if $\vec{x} = 0$, so

$$(\vec{x}, \vec{x}) \geq 0; \quad \text{if} \quad (\vec{x}, \vec{x}) = 0, \quad \text{then} \quad \vec{x} = 0.$$

The length (or magnitude) of a vector $\vec{x}$ is written as $|\vec{x}|$, so

$$|\vec{x}|^2 = (\vec{x}, \vec{x}).$$

It turns out that if we know the length, defined in this way, of any vector of space A^n, we can calculate the scalar product of any two vectors. Because of symmetry and linearity we have

$$(\vec{x} + \vec{y}, \vec{x} + \vec{y}) = (\vec{x}, \vec{x}) + 2(\vec{x}, \vec{y}) + (\vec{y}, \vec{y}).$$

Therefore, we get

$$(\vec{x}, \vec{y}) = \frac{1}{2}(|\vec{x} + \vec{y}|^2 - |\vec{x}|^2 - |\vec{y}|^2). \tag{4.20}$$

Here on the left side is a scalar product $(\vec{x}, \vec{y})$ of two vectors of the space A^n, and on the right side are lengths of three vectors that are known according to our assumption.

B) A set of vectors

$$\vec{\varepsilon}_1, \vec{\varepsilon}_2, ..., \tag{4.21}$$

is called orthonormal if the following relation occurs

$$(\vec{\varepsilon}_i, \vec{\varepsilon}_j) = \begin{cases} 1, & i = j; \\ 0, & i \neq j; \end{cases} \quad i, j = 1, ..., p \tag{4.22}$$

Therefore, two different vectors of the set (4.21) are orthogonal to each other, and the length of any vector of (4.21) is equal to one.

The orthonormal set of vectors (4.21) is always linearly independent.

Indeed, if

$$\alpha_1 \vec{\varepsilon}_1 + \alpha_2 \vec{\varepsilon}_2 + ... + \alpha_p \vec{\varepsilon}_p = 0,$$

then, applying scalar multiplication by ε_i, we get $\alpha_i = 0$ for all i.

Thus, if $p = n$, then set (4.21) is an orthonormal basis

$$\vec{\varepsilon}_1, \vec{\varepsilon}_2, ..., \vec{\varepsilon}_n \qquad (4.23)$$

of the space A^n.

In the case of an orthonormal basis (4.23), the coordinate notation of a scalar product (4.19) can be written as

$$(\vec{x}, \vec{y}) = \sum_{i=1}^{n} x^i y^i.$$

A vector basis (4.23) that complies with condition (4.22) is called orthonormal because every two nonidentical vectors of this set are orthogonal, and the basis is normalized because the length of each vector is equal to one.

C) Every linearly independent set of vectors

$$\vec{x}_1, \vec{x}_2, ..., \vec{x}_p \qquad (4.24)$$

defines the single, orthonormal set of vectors

$$\vec{\varepsilon}_1, \vec{\varepsilon}_2, ..., \vec{\varepsilon}_p \qquad (4.25)$$

The process of transforming set (4.24) into set (4.25) is called ortho-normalization. Let us describe the process.

Since set (4.24) is linearly independent, the vector $\vec{x}_1 \neq 0$, and therefore, its length $|\vec{x}_1| \neq 0$. Let us denote

$$\vec{\varepsilon}_1 = \frac{\vec{x}_1}{|\vec{x}_1|}$$

Then, let us denote

$$\vec{\varepsilon}_2' = \vec{x}_2 - (\vec{x}_2, \vec{\varepsilon}_1)\vec{\varepsilon}_1$$

We get

$$(\vec{\varepsilon}_2', \vec{\varepsilon}_1) = (\vec{x}_2, \vec{\varepsilon}_1) - (\vec{x}_2, \vec{\varepsilon}_1) = 0$$

Therefore, vectors $\vec{\varepsilon}_1$ and $\vec{\varepsilon}_2'$ are orthogonal to each other. The vector $\vec{\varepsilon}_2' \neq 0$ because the vectors $\vec{x}_1$ and $\vec{x}_2$ are linearly independent, and therefore, the vectors $\vec{\varepsilon}_1$ and $\vec{\varepsilon}_2'$ are also linearly independent. Now let us normalize the $\vec{\varepsilon}_2'$ vector. Let us denote

$$\vec{\varepsilon}_2 = \frac{\vec{\varepsilon}_2'}{|\vec{\varepsilon}_2'|}.$$

Therefore, we get

$$(\vec{\varepsilon}_2, \vec{\varepsilon}_2) = 1.$$

Let us suppose that vectors $\vec{\varepsilon}_1, \vec{\varepsilon}_2, ..., \vec{\varepsilon}_i$ are already made in such a way that they constitute some orthonormal set. Let us produce the $\vec{\varepsilon}_{i+1}'$ vector. Let us denote $\vec{\varepsilon}_{i+1}'$ as

$$\vec{\varepsilon}_{i+1}' = \vec{x}_{i+1} - (\vec{x}_{i+1}, \vec{\varepsilon}_1)\vec{\varepsilon}_1 - (\vec{x}_{i+1}, \vec{\varepsilon}_2)\vec{\varepsilon}_2 - ... - (\vec{x}_{i+1}, \vec{\varepsilon}_i)\vec{\varepsilon}_i.$$

First of all, vector $\vec{\varepsilon}_{i+1}' \neq 0$, because from linear independence of the vectors $\vec{x}_1, \vec{x}_2, ..., \vec{x}_{i+1}$ follows linear independence of the vectors

$$\vec{\varepsilon}_1, \vec{\varepsilon}_2, ..., \vec{\varepsilon}_i, \vec{x}_{i+1}.$$

Multiplying the vector $\vec{\varepsilon}_{i+1}'$ by any vector $\vec{\varepsilon}_j, j \leq i$ we get

$$(\vec{\varepsilon}_{i+1}', \vec{\varepsilon}_j) = 0.$$

Therefore, the vector $\vec{\varepsilon}'_{i+1}$ is orthogonal to all vectors $\vec{\varepsilon}_1, \vec{\varepsilon}_2,$ ..., $\vec{\varepsilon}_i$, and the vector is not equal to zero. Normalizing the vector $\vec{\varepsilon}'_{i+1}$, in other words, setting

$$\vec{\varepsilon}_{i+1} = \frac{\vec{\varepsilon}'_{i+1}}{|\vec{\varepsilon}'_{i+1}|},$$

we get the vector $\vec{\varepsilon}_{i+1}$. In this way we have finished the inductive process of making of an orthonormal set (4.25). Note that conversion of set (4.24) into set (4.25) was contiguous. That means that in the process of changing set (4.24) contiguously, set (4.25) was also changed contiguously.

D) Each vector subspace B^p of a Euclidean vector space A^n is, naturally, itself a Euclidean vector space.

Indeed, every pair of vectors from B^p belongs to A^n, and therefore, the scalar product operation is defined for them. A vector z of A^n which is orthogonal to each vector of B^p is regarded as orthogonal to the whole subspace B^p. Let us denote as C a set of all the vectors of space A^n orthogonal to the subspace B^p. It turns out that C is a subspace of A^n of dimension $q = n - p$, and the space A^n is the Cartesian product of its subspaces B^p and $C = C^q$.

The space C^q is called an orthogonal complement of the space B^p.

It turns out that the subspace B^p is an orthogonal complement of subspace C^q.

In order to prove this statement let us choose in subspace B^p some orthonormal basis

$$\vec{\varepsilon}_1, \vec{\varepsilon}_2, ..., \vec{\varepsilon}_p.$$

Now let us expand this set of vectors to the maximum, linearly independent set in A^n by adding some vectors $\vec{x}_{p+1}, ..., \vec{x}_n$ (see Section 4.1, statement C)). We get a set of linearly independent vectors

$$\vec{\varepsilon}_1, \vec{\varepsilon}_2, ..., \vec{\varepsilon}_p, \vec{x}_{p+1}, ..., \vec{x}_n.$$

Let us ortho-normalize it (see statement C)). The first p vectors of the set will not change, and the vectors $\vec{x}_{p+1}, \vec{x}_{p+2}, ..., x_n$ will be replaced. Let us write the resulting set as

$$\vec{\varepsilon}_1, \vec{\varepsilon}_2, ..., \vec{\varepsilon}_p, \vec{\varepsilon}_{p+1}, ..., \vec{\varepsilon}_n.$$

This set is an orthonormal basis of the space A^n, so each vector of A^n may be written as

$$\vec{x} = x^1 \vec{\varepsilon}_1 + ... + x^p \vec{\varepsilon}_p + x^{p+1} \vec{\varepsilon}_{p+1} + ... + x^n \vec{\varepsilon}_n.$$

In order for vector x to be orthogonal to all of B^p, it has to be orthogonal to each vector $\vec{\varepsilon}_i, i = 1, ..., p$. Therefore, it should comply with the following condition

$$(\vec{x}, \vec{\varepsilon}_i) = 0; \ i = 1, ..., p.$$

From this it follows that

$$x^1 = 0, ..., x^p = 0.$$

Therefore, vector z that is orthogonal to the vector space B^p may be written in the following way

$$\vec{z} = x^{p+1} \vec{\varepsilon}_{p+1} + ... + x^n \vec{\varepsilon}_n.$$

A set of all such vectors constitutes the vector space C^q, a subspace of A^n, with $q = n - p$ dimensions, so it is an orthogonal complement of the subspace B^p.

It is clear that the subspace B^p in turn is also an orthogonal complement of the subspace C^q.

Let us give a geometric interpretation of the scalar product of two vectors of A^n.

E) Let x, y be two linearly independent vectors of A^n. The set of all vectors

$$\vec{u} = \alpha \vec{x} + \beta \vec{y},$$

such that α and β are real numbers, constitutes a two dimensional subspace A^2 of the space A^n. The space A^2 is a two dimensional Euclidean vector space, in other words, a Euclidean plane. Therefore, vectors $\vec{x}$ and $\vec{y}$ are two vectors of the plane A^2, and we can determine what the angle between these two vectors is. Let us name it φ. It turns out that the scalar product of two vectors $\vec{x}$ and $\vec{y}$ is

$$(\vec{x}, \vec{y}) = |\vec{x}||\vec{y}| \cos \varphi. \tag{4.26}$$

To prove this let us orthonormalize the vectors $\vec{x}$ and $\vec{y}$. We get two vectors

$$\vec{\varepsilon}_1, \vec{\varepsilon}_2,$$

that constitute a basis of the two dimensional Euclidean space A^2, and

$$\vec{x} = |\vec{x}|\vec{\varepsilon}_1; \quad \vec{y} = |\vec{y}|(\vec{\varepsilon}_1 \cos \varphi + \vec{\varepsilon}_2 \sin \varphi). \tag{4.27}$$

Therefore,

$$(\vec{x}, \vec{y}) = (|\vec{x}|\vec{\varepsilon}_1, |\vec{y}|(\vec{\varepsilon}_1 \cos \varphi + \vec{\varepsilon}_2 \sin \varphi)) = |\vec{x}||\vec{y}| \cos \varphi.$$

We have proved formula (4.26). ■

If $\vec{x}$ and $\vec{y}$ are linearly dependent, then the angle between them is either equal to zero, or to π. In that case both of them can be expressed using the same vector:

$$x = |\vec{x}|\vec{\varepsilon}_1; \quad \vec{y} = \pm|\vec{y}|\vec{\varepsilon}_1.$$

Then we have

$$(\vec{x}, \vec{y}) = |\vec{x}||\vec{y}| \cos 0$$

or

$$(\vec{x}, \vec{y}) = |\vec{x}||\vec{y}| \cos \pi.$$

Therefore, formula (4.26) is correct for any two vectors $\vec{x}$ and $\vec{y}$ of A^n.

F) Let A^3 be a Euclidean vector space of three dimensions, and

$$\vec{\varepsilon}_1, \vec{\varepsilon}_2, \vec{\varepsilon}_3 \qquad (4.28)$$

be its orthonormal basis. Let us define the vector product

$$\vec{z} = [\vec{x}, \vec{y}]$$

of two vectors

$$\vec{x} = x^1 \vec{\varepsilon}_1 + x^2 \vec{\varepsilon}_2 + x^3 \vec{\varepsilon}_3,$$

$$\vec{y} = y^1 \vec{\varepsilon}_1 + y^2 \vec{\varepsilon}_2 + y^3 \vec{\varepsilon}_3.$$

Let us define coordinates z^1, z^2, z^3 of vector $\vec{z}$ in basis 4.28 using the following formulas:

$$z^1 = x^2 y^3 - x^3 y^2,$$

$$z^2 = x^3 y^1 - x^1 y^3, \qquad (4.29)$$

$$z^3 = x^1 y^2 - x^2 y^1.$$

Later on we will discuss whether this definition is invariant with respect to our choice of basis (4.28) (see Section 4.4). For now, let us only note that

$$[\vec{x}, \vec{y}] = -[\vec{y}, \vec{x}]; \quad [\vec{x}, \vec{x}] = 0.$$

Let us calculate the coordinates of $\vec{z}$ using formulas (4.27). To do so, let us choose an orthonormal basis of A^3 in a special way, so that vectors $\vec{x}$ and $\vec{y}$ can be written using formulas (4.27), and then let us calculate the vector product $\vec{z}$ using these formulas. From these formulas and formula (4.29) follows:

$$z^1 = 0, \ z^2 = 0, \ z^3 = |\vec{x}||\vec{y}| \sin \varphi,$$

where φ is the angle between vectors $\vec{x}$ and $\vec{y}$. Therefore, the vector product of vectors $\vec{x}$ and $\vec{y}$ is orthogonal to them, and its modulus is equal to the product of modulus $|\vec{x}|$, modulus $|\vec{y}|$, and the sine, and its direction is defined by the angle φ.

G) A mapping φ of a Euclidean vector space A^n onto a Euclidean space $\hat{A}^n$ is isomorphic if it is linear (see Section 4.1 G)) and preserves the value of scalar products. In other words, for any vectors $\vec{x}$ and $\vec{y}$ of A^n, the following formula is true

$$(\varphi(\vec{x}), \varphi(\vec{y})) = (\vec{x}, \vec{y}). \tag{4.30}$$

Let us note that requiring formula (4.20) to preserve the scalar product follows from the property of preserving the length of a vector in mapping φ. Specifically, if for any vector $x \in A^n$ the following relation is true

$$|\varphi(\vec{x})| = |\vec{x}|,$$

then the mapping φ also preserves the scalar product (see (4.30)). Let φ_t be an isomorphic mapping of Euclidean space A^n onto itself that depends continuously on parameter t, which changes within the interval $0 \le t \le 1$, and φ_0 is the identity mapping of A^n onto itself, in other words

$$\varphi_0(\vec{x}) = \vec{x}.$$

Therefore, the mapping φ_1 is derived from the identity mapping by continuous modification of it; in other words, by continuous deformation.

An isomorphic mapping of a Euclidean vector space onto itself that is made by continuous deformation of the identity mapping is called a rotation of Euclidean vector space A^n.

Let us note here that not all isomorphic mappings of a Euclidean vector space onto itself are rotations. It will be proved in Section 4.4.

In Euclidean vector space we can define a distance $\rho(\vec{x}, \vec{y})$ between any points $\vec{x}$ and $\vec{y}$, or equivalently, between any two vectors $\vec{x}$ and $\vec{y}$ that start from the origin and point to the points $\vec{x}$ and $\vec{y}$ respectively. The distance is defined by the following formula

$$\rho(\vec{x}, \vec{y}) = |\vec{x} - \vec{y}|. \tag{4.31}$$

4.3 Quaternions

The complex numbers play a huge role in mathematics. This was the reason for attempts to further generalize the real numbers. On the way to finding possible further generalizations, the quaternions were constructed. It has turned out that their role in mathematics is insignificant.

The quaternions are the result of adding to real numbers not one but three imaginary components, that are named as

$$i, j, k, \tag{4.32}$$

so that each quaternion can be written as

$$x = x^0 + x^1 i + x^2 j + x^3 k, \tag{4.33}$$

where

$$x^0, x^1, x^2, x^3$$

are real numbers that are the coordinates of the quaternion x. Therefore, the set of all quaternions K^4 is a four-dimensional vector space with the basis

$$1, i, j, k.$$

If this basis is orthonormal, then the vector space K^4 is an Euclidean four-dimensional vector space.

A sum of quaternions x and y, where

$$x = x^0 + x^1 i + x^2 j + x^3 k,$$

$$y = y^0 + y^1 i + y^2 j + y^3 k$$

is defined as the sum of vectors, in other words, it is defined by the following formula

$$x + y = x^0 + y^0 + (x^1 + y^1)i + (x^2 + y^2)j + (x^3 + y^3)k.$$

A quaternion x becomes a real number x^0, if its coordinates x^1, x^2, x^3 are equal to zero. Therefore, the set K^4 of all quaternions includes a real axis D of all real quaternions. The real number x^0 is the real part of quaternion (4.33). All quaternions x that have $x^0 = 0$ are purely imaginary. Those quaternions constitute a three dimensional subspace I of the space K^4. The spaces I and D are orthogonal complements of each other. Therefore, quaternion x may be written as

$$x = x^0 + \hat{x}$$

where $\hat{x} = x^1 i + x^2 j + x^3 k$. Here x^0 is the real part of the quaternion, and $\hat{x}$ is the imaginary part. Let us write quaternion y in the same way

$$y = y^0 + \hat{y},$$

where

$$\hat{y} = y^1 i + y^2 j + y^3 k.$$

The conjugate quaternion $\bar{x}$, produced from x, is defined by the following formula

$$\bar{x} = x^0 - \hat{x}$$

Analogously

$$\bar{y} = y^0 - \hat{y}.$$

Let us move on now to the description of the rules of quaternion multiplication. Let us regard multiplication of a quaternion coordinate, which is a real number, by one of the imaginary elements of a quaternion (4.32) as commutative. After that we only need to specify the rules of multiplication of the imaginary elements (4.32).

These rules are:

$$i^2 = j^2 = k^2 = -1, \tag{4.34}$$

$$ij = -ji = k; jk = -kj = i;$$
$$ki = -ik = j \tag{4.35}$$

Using these rules let us first multiply the imaginary parts of quaternions $\hat{x}$ and $\hat{y}$. We have:

$$\hat{x}\hat{y} = -(x^1y^1 + x^2y^2 + x^3y^3) + (x^2y^3 - x^3y^2)i + (x^1y^3 - x^3y^1)j + (x^1y^2 - x^2y^1)k.$$

Using the rules of scalar and vector multiplication (see Section 4.2 B), F)) of vectors $\hat{x}$ and $\hat{y}$ of three dimensional Euclidean vector space I, we can rewrite this formula as follows:

$$\hat{x}\hat{y} = -(\hat{x}, \hat{y}) + [\hat{x}, \hat{y}] \tag{4.36}$$

Using this formula we can write the product of two quaternions x and y in the following way

$$xy = x^0y^0 - (\hat{x}, \hat{y}) + x^0\hat{y} + y^0\hat{x} + [\hat{x}, \hat{y}]. \tag{4.37}$$

Let us now calculate the product of quaternion x and quaternion $\bar{x}$. We have

$$x\bar{x} = (x^0)^2 + (\hat{x}, \hat{x}) = (x, x) = |x|^2. \tag{4.38}$$

Therefore, the product of $x\bar{x}$ is the scalar power of two of vector x in the Euclidean vector space K^4.

From this it follows that each quaternion x, not equal to zero, has an inverse quaternion x^{-1}, that satisfies the following condition $xx^{-1} = x^{-1}x = 1$, and

$$x^{-1} = \frac{\bar{x}}{|x|}. \tag{4.39}$$

Let us write the product of quaternions $\bar{y}$ and $\bar{x}$ using formula (4.37). We have

$$\bar{y}\bar{x} = x^0 y^0 - (\hat{x}, \hat{y}) - x^0 \hat{y} - y^0 \hat{x} - [\hat{x}, \hat{y}].$$

From this formula we get the following important formula

$$\overline{xy} = \bar{y}\bar{x}. \tag{4.40}$$

Let us prove now that

$$|xy| = |x| \cdot |y|. \tag{4.41}$$

Indeed, using formula (4.38), we have

$$|xy|^2 = xy\overline{xy} = xy\bar{y}\bar{x} = x|y|^2\bar{x} = x\bar{x}|y|^2 = |x|^2|y|^2.$$

From formula (4.38) it follows that if the modulus of a quaternion ε is equal to 1, then

$$\varepsilon^{-1} = \bar{\varepsilon}. \tag{4.42}$$

Let us call H the set of all quaternions with modulus equal to 1. The set H is a three-dimensional sphere in the K^4 space, with radius equal to 1 and center at the origin.

Let us note that each quaternion a with modulus equal to 1 can be written as follows

$$a = \cos\alpha + (\sin\alpha)u, \tag{4.43}$$

63

where α is an angle such that $|\alpha| \leq \pi$ and u is a pure imaginary quaternion with modulus equal to 1, and visa versa. Every quaternion that may be written in this form has its modulus equal to one.

Let us prove formula (4.43). Since the space K^4 is a Cartesian product of its subspaces D and I, then

$$a = \xi + \eta u$$

where ξ and η are real numbers, and u is a pure imaginary quaternion, with modulus equal to 1. Then,

$$1 = |a|^2 = a\bar{a} = (\xi + \eta u)(\xi - \eta u) = \xi^2 - \eta^2 u^2 = \xi^2 + \eta^2.$$

Therefore, $\xi^2 + \eta^2 = 1$, and there is an angle α, $|\alpha| \leq \pi$, such that $\xi = \cos\alpha$, $\eta = \sin\alpha$. Thus formula (4.43) is proved. ∎

Taking into account that quaternions are added in the same way as vectors, we have

$$|x + y| \leq |x| + |y|. \tag{4.44}$$

It turns out that the operations of quaternion addition and multiplication are contiguous. That means that if a quaternion x' differs only slightly from the quaternion x, and a quaternion y' differs only slightly from the quaternion y, then in turn quaternion $x' + y'$ differs only slightly from the quaternion $x + y$, and the quaternion $x'y'$ differs only slightly from the quaternion xy.

Let us prove that. We have

$$(x' + y') - (x + y) = (x' - x) + (y' - y).$$

Because according to our supposition, $x' - x$, and $y' - y$ are small quaternions, then from formula 4.44 it follows that quaternion $(x' + y') - (x + y)$ is also small.

Let us move on to multiplication. We have

$$x^{'} y^{'} - xy = x^{'} y^{'} - x^{'} y + x^{'} y - xy =$$
$$x^{'} (y^{'} - y) + (x^{'} - x)y.$$

Because $y^{'} - y$ and $x^{'} - x$ are small quaternions, from formula (4.41) it follows that quaternions $x^{'} (y^{'} - y)$ and $(x^{'} - x)y$ are also small. From formula (4.44) it follows that their sum is also small.

4.4 Geometric applications of quaternions

In the previous section we built the set of quaternions K^4, which is a four dimensional Euclidean vector space. We have found that this space is the Cartesian product of D, the set of real quaternions, and I, the set of pure imaginary quaternions, and that the linear subspaces D and I of the space K^4 are orthogonal complements of each other.

In the space K^4 we selected a set H of quaternions such that the modulus of any quaternion from this set is equal to one. This set is a sphere with radius 1 and its center at the origin.

A) Let us assign to each quaternion a from the set H a mapping f_a of the Euclidean vector space K^4 onto itself using the following formula

$$f_a(x) = axa^{-1}. \tag{4.45}$$

It turns out that the mapping f_a is isomorphic; in other words it preserves all operations that exist in the set K^4. In particular, the following relations are true

$$f_a f_b = f_{ab}.$$

It is easy to check that successive application of two such mappings f_a and f_b produces a mapping that is described by the following formula

$$f_a(x + y) = f_a(x) + f_a(y), \tag{4.46}$$

$$f_a(xy) = f_a(x)f_a(y).$$

Furthermore, it turns out that the linear mapping f_a of the Euclidean vector space K^4 is a rotation (see Section 4.2 G)). Moreover, the mapping f_a maps each vector subspace D or I into itself.

Let us prove statement A). First of all let us prove relation (4.46). The proof is performed automatically. Indeed we have

$$f_a(x + y) = a(x + y)a^{-1} = axa^{-1} + aya^{-1} = \\ f_a(x) + f_a(y).$$

Furthermore,

$$f_a(xy) = axya^{-1} = axa^{-1}aya^{-1} = f_a(x)f_a(y).$$

Let us prove now that mapping f_a preserves the length of a vector x. First of all let us note that

$$\overline{axa^{-1}} = \bar{a}^{-1}\bar{x}\bar{a} = a\bar{x}a^{-1}$$

(See formulas (4.40), and (4.42)).

The length of the vector axa^{-1} can be defined, using formula 4.38, as

$$|axa^{-1}|^2 = axa^{-1}\overline{axa^{-1}} = axa^{-1}a\bar{x}a^{-1} = \\ ax\bar{x}a^{-1} = |x|^2.$$

As we proved in Section 4.2, from the fact that the mapping f_a preserves the length of a vector it follows that the scalar product, defined in the Euclidean vector space K^4, is also preserved by the mapping f_a. Therefore, mapping f_a is an isomorphic mapping of the three dimensional vector space I onto itself.

Furthermore, we have for x from D

$$axa^{-1} = xaa^{-1} = x.$$

Therefore, $f_a(x) = x$, the mapping f_a of D onto itself is the identity mapping. The orthogonal complement I of D is also mapped into itself, because the f_a mapping preserves a vector's orthogonality.

Let us prove now that the mapping f_a is a rotation of the Euclidean vector space K^4. Using formula (4.43) we have

$$a = \cos\alpha + (\sin\alpha)u.$$

If we define $a(t) = \cos\alpha t + (\sin\alpha t)u$, we get a quaternion $a(t)$, with modulus equal to one, and while t is changing from 0 to 1, it changes from 1 to a. Therefore, the mapping $f_{a(t)}$ makes a continuous transformation of the mapping f_1 into the mapping f_a; therefore, f_a is a rotation of the Euclidean space K^4.

B) Let u, v be an orthonormal pair of vectors from I. Consider them as quaternions, and let us reckon the result of their multiplication

$$w = uv.$$

It turns out that for the quaternions

$$u, v, w \qquad (4.47)$$

the same relationships hold true that were originally established for the pure imaginary units of quaternions 4.32. So the following relationship holds

$$u^2 = v^2 = w^2 = -1 \qquad (4.48)$$

Furthermore,

$$uv = -vu = w$$
$$vw = -wv = u \qquad (4.49)$$
$$wu = -uw = v$$

(compare with (4.34), (4.35)).

67

Let us prove relationships (4.48) and (4.49). Using formula (4.36) we have

$$uu = -(u, u) + [u, u] = -1$$

In the same way

$$vv = -(v, v) + [v, v] = -1$$

Using the same formula (4.36) we have

$$uv = [u, v] = -[v, u] = -vu.$$

Thus the quaternion $w = uv$ is defined in the three dimensional Euclidean space I as a vector multiplication of vectors u, v, and therefore, belongs to I. Also the following relationship is takes place

$$uv = -vu = w.$$

We have proved, therefore, the first relationship (4.49). Furthermore, we have

$$(uv)^2 = uvuv = -uuv \cdot v = -1$$

that proves the last of relationships (4.48). Let us prove now the last two relationships (4.49). We have

$$vw = vuv = -v^2 u = u,$$
$$wu = uvu = -vu^2 = v.$$

Therefore, all the relationships (4.49) are proved. ∎

C) Let u, v be an arbitrary orthonormal pair from I. Let us denote the result of multiplication of u by v as w, i.e. $w = uv$.

Let us denote by P a plane from I that has u and v as its basis. Let a be

$$a = \cos\alpha + \sin\alpha \cdot u, \quad |\alpha| \le \pi$$

(See (4.43).) It turns out that the rotation f_a (see (4.45)) of space I is a rotation around the axis u, such that the plane P rotates in the direction from vector v to vector w about the angle 2α. The angle α may be negative, and in that case the rotation goes in the opposite direction.

Let us prove statement C). Let us calculate the following quaternions

$$f_a(u) = aua^{-1},$$
$$f_a(v) = ava^{-1},$$
$$f_a(w) = awa^{-1}.$$

Since the quaternion a is equal to $\cos\alpha + \sin\alpha \cdot u$, multiplication of a by u is commutative, and therefore, we have:

$$f_a(u) = aua^{-1} = uaa^{-1} = u.$$

Furthermore, we have

$$f_a(v) = (\cos\alpha + \sin\alpha \cdot u)v(\cos\alpha - \sin\alpha \cdot u) =$$
$$= (\cos\alpha + \sin\alpha \cdot u)(\cos\alpha + \sin\alpha \cdot u)v =$$
$$= (\cos 2\alpha + \sin 2\alpha \cdot u)v =$$
$$= \cos 2\alpha \cdot v + \sin 2\alpha \cdot w.$$

Furthermore,

$$f_a(w) = (\cos\alpha + \sin\alpha \cdot u)w(\cos\alpha - \sin\alpha \cdot u) =$$
$$= (\cos\alpha + \sin\alpha \cdot u)(\cos\alpha + \sin\alpha \cdot u)w =$$
$$= (\cos 2\alpha + \sin 2\alpha \cdot u)w =$$
$$= \cos 2\alpha \cdot w - \sin 2\alpha \cdot v.$$

From the last two formulas it follows that the mapping f_a produces a rotation of plane P in the direction from the vector v to the vector w about the angle 2α.

D) Let x and y be two quaternions from I, and let their moduli both be equal to one. Then there exists a rotation f_a, such that

$$f_a(x) = y.$$

In other words, there is a rotation of the space I that maps the vector x into the vector y, and that can be named f_a. If x is very close to y, then the angle α is small, and therefore, quaternion a is close to one.

In order to prove the statement D), let us draw a plane P that includes vectors x and y; let u be a quaternion from I orthogonal to the plane P, and let the modulus of u be equal to one. Let us choose an arbitrary vector v of the plane P, such that $|v| = 1$. Then u and v are an orthonormal pair in I. Let w be

$$w = uv,$$

then we get an orthonormal basis (4.47) of the space I, and the orthonormal pair v and w is a basis of the plane P that vectors x and y belong to. It is clear that there is an angle α such that the rotation through the angle 2α in the direction from v to w will transform the vector x to the vector y, and α might be also negative. If the vectors x and y are close to each other, then the angle α will be small. Thus, the statement D) is proven. ∎

E) There exists a rotation f_c(see A)), such that

$$f_c(i) = u,$$
$$f_c(j) = v,$$
$$f_c(k) = w,$$

where u, v, w are the quaternions constructed in B). If the orthonormal system i, j, k (4.32) is close to the orthonormal system u, v, w (4.40), then the quaternion c is close to one, and the rotation f_c is small.

Let us prove statement E). Statement D) says that there is a rotation f_a that maps the quaternion i to u, in other words

$$f_a(i) = u.$$

Also if the orthonormal systems (4.32) and (4.40) are close, then the quaternion a is close to one. Let P be a plane with the orthonormal basis j, k, and

$$P' = f_a(P),$$

then the plane P' includes quaternions $f_a(j)$, $f_a(k)$, v, and w. Now, there is a rotation f_b, $b = \sin\beta + \cos\beta \cdot u$, around the axis u that transforms quaternion $f_a(j)$ into quaternion v; and if the orthonormal systems (4.32) and (4.47) are close to each other, then the quaternion b is close to one.

Therefore, we have

$$f_b(f_a(j)) = v.$$

If we let c be $c = ba$, then we get

$$f_c(i) = u, f_c(j) = v.$$

Furthermore,

$$f_c(k) = f_c(ij) = f_c(i)f_c(j) = uv = w,$$

and if the orthonormal systems (4.32) and (4.47) are close to each other, then quaternions a and b are close to one, and therefore, the quaternion c is also close to one.

F) Let

$$u_0, v_0, w_0 \qquad \text{and} \qquad u_1, v_1, w_1$$

be two orthonormal triplets in the space I that are close to each other. Then there is a rotation f_c, where c is a quaternion that is close to one, such as

$$f_c(u_0) = u_1,$$
$$f_c(v_0) = v_1, \qquad\qquad (4.50)$$
$$f_c(w_0) = w_1.$$

Let us prove statement F). Because the quaternion $w_0' = u_0 v_0$ is orthogonal to the quaternions u_0, v_0, we have

$$w_0' = \varepsilon_0 w_0, \qquad \text{where} \qquad \varepsilon_0 = \pm 1.$$

Because the quaternion $w_1' = u_1 v_1$ is orthogonal to the quaternions u_1, v_1, we have

$$w_1' = \varepsilon_1 w_1, \qquad \text{where} \qquad \varepsilon_1 = \pm 1.$$

Because quaternion pairs u_0, v_0 and u_1, v_1 are close to each other, the products of their multiplication w_0' and w_1' are also close to each other. From the closeness of w_0' and w_1' it follows that $\varepsilon_0 = \varepsilon_1 = \varepsilon$, where $\varepsilon = \pm 1$.

The quaternion triplets

$$u_0, v_0, \varepsilon w_0 \qquad \text{and} \qquad u_1, v_1, \varepsilon w_1,$$

are close to each other and constitute an orthonormal system in I. They can be used as the imaginary ones of the system K^4 (see statement B)). Therefore, taking into account statement E), there exists a rotation f_c, where c is a quaternion close to one, such that

$$f_c(u_0) = u_1,$$
$$f_c(v_0) = v_1,$$
$$f_c(\varepsilon w_0) = \varepsilon w_1.$$

Therefore, relationships (4.50) hold. The statement E) has been proved. ∎

Theorem 1 (Rotations of pure imaginary quaternions):
Every rotation g of the vector space I composed of all pure imaginary quaternions can be written in the form

$$g(x) = f_a(x) = axa^{-1},$$

where a is a quaternion with modulus equal to one. If the moduli of two quaternions a and b are equal to one, then rotations f_a and f_b are identical, if and only if, a is identical to b, or a and b differ only by sign, i.e.

$$b = \pm a.$$

Because g is a rotation, it is produced as a result of some continuous transformation φ_t of the identity mapping φ_0 into the rotation $\varphi_1 = g$ (see Section 4.2, G)). Here $0 \leq t \leq 1$, and φ_t is a rotation that continuously depends on the parameter t. Let us split the interval $[0,1]$ into n equal parts of length δ, and δ is so small that the rotations $\varphi_{p\delta}$ and $\varphi_{(p+1)\delta}$ differ very insignificantly one from another. This means that for any vector x of I with length equal to one, quaternions $\varphi_{p\delta}(x)$ and $\varphi_{(p+1)\delta}(x)$ are close to each other.

Let i, j, k be quaternion imaginary ones of I (4.32). Then, because the orthonormal triplets

$$\varphi_{p\delta}(i), \varphi_{p\delta}(j), \varphi_{p\delta}(k);$$
$$\varphi_{(p+1)\delta}(i), \varphi_{(p+1)\delta}(j), \varphi_{(p+1)\delta}(i)$$

are close to each other, according to Statement F) there is a quaternion c_p close to one, such that rotation f_{c_p} transforms the first triplet into the second one (see F)). Thus sequential application of rotations $f_{c_1}, f_{c_2}, ... f_{c_n}$ will transform the orthonormal triplet i, j, k into the triplet $g(i), g(j), g(k)$.

Therefore, the rotation g will be represented as a sequence of rotations $f_{c_1}, ..., f_{c_n}$, so the quaternion a we are looking for may be represented by the formula

$$a = c_n c_{n-1} ... c_1$$

Therefore, the first part of the theorem has been proven. ∎

Let us prove now that rotations f_a and f_b are identical if and only if $a = \pm b$.

If this equality takes place, then it is clear that rotations f_a and f_b are identical. Let us prove the reverse statement. Let us consider the rotation $f_c = f_{a^{-1}}f_b$. In that case $c = a^{-1}b$. If the rotations f_a and f_b are identical, then the rotation f_c is the identity. Let

$$c = \cos\alpha + \sin\alpha \cdot u, \ |\alpha| \leq \pi$$

(see Statement D)). Then, due to Statement B), rotation f_c is the rotation around the axis u through the angel 2α. Such a rotation is the identity rotation only on the condition that $2\alpha = 2q\pi$, in other words $\alpha = q\pi$. In that case $c = \pm 1$, and therefore, $a = \pm b$. Therefore, Theorem 1 has been proven. ■

Let us look now at the rotation of four dimensional Euclidean space K^4 that comprises all quaternions.

G) Let us associate with each pair of quaternions a and b $\in H$, with moduli equal to one, a mapping

$$f_{a,b}(x) = axb^{-1},$$

where x is an arbitrary vector of K^4, i.e. an arbitrary quaternion. It turns out that $f_{a,b}$ is a rotation of the Euclidean vector space K^4.

In order to prove this, let us first prove that this mapping does not change the length of the vector, i.e. the modulus of the quaternion x. Using formula (4.41), we have

$$|axb^{-1}| = |a| \cdot |x| \cdot |b^{-1}| = |x|.$$

Therefore, the moduli of quaternions x and $f_{a,b}(x)$ are equal, and consequently due to formula (4.20), $f_{a,b}$ is an isomorphic mapping of the Euclidean vector space K^4 into itself.

Let us prove now that an isomorphic mapping $f_{a,b}$ of the Euclidean vector space K^4 into itself is a rotation, i.e. that it is a product of continuous transformation φ_t, with $0 \leq t \leq 1$, of the identity mapping φ_0 into the mapping $\varphi_{a,b}$. To that end, let us rewrite quaternions a and b in the form D):

$$a = \cos \alpha + \sin \alpha \cdot u,$$
$$b = \cos \beta + \sin \beta \cdot v.$$

Furthermore,

$$a(t) = \cos \alpha t + \sin \alpha t \cdot u,$$
$$b(t) = \cos \beta t + \sin \beta t \cdot v.$$

The quaternions $a(t)$ and $b(t)$ depend continuously on the parameter t, and their moduli are equal to 1. Therefore, the mapping $f_{a(t),b(t)}$ is an isomorphic mapping of the Euclidean vector space K^4 into itself, which depends continuously on t, and transforms the identity mapping into the mapping $f_{a,b}$. Therefore, $f_{a,b}$ is a rotation of the Euclidean vector space K^4.

Theorem 2 (Rotation of Euclidean vector space):
Every rotation g of the Euclidean vector space K^4 can be given by the formula

$$g(x) = f_{a,b}(x) = axb^{-1} \qquad (4.51)$$

(See statement G)) The rotations $f_{a,b}$ and $f_{a',b'}$ are identical if and only if quaternions a' and b' are identical with quaternions a and b respectively, or they differ only in sign.

Let ε be

$$\varepsilon = g(1) \qquad (4.52)$$

Here ε is a quaternion with modulus equal to 1. Alongside the mapping g let us consider the mapping defined by the following formula

$$g' = \varepsilon^{-1} g \qquad (4.53)$$

It is easy to check that the mapping g' is a rotation of the Euclidean vector space K^4, because the quaternion ε can be gotten from the quaternion unit by way of the continuous modification $\varepsilon(t)$, where $|\varepsilon(t)| = 1$, that transforms the quaternion one into the quaternion ε. In addition, multiplication of all quaternions of K^4 by quaternion $\varepsilon(t)$, which has modulus equal to one, is an isomorphic mapping of the Euclidean vector space K^4 into itself (see statement F)). Therefore, g' is a rotation of the Euclidean vector space K^4. From the formulas (4.52), (4.53) it follows that the whole real axis D in the Euclidean vector space K^4 is mapped identically into itself by the mapping g'. Because the mapping g' is an isomorphic mapping of the Euclidean vector space K^4 into itself, then the orthogonal compliment to D, namely I, is mapped into itself by g'. Therefore, g' is a rotation of the Euclidean vector space I.

So according to Theorem 1, it may be written as

$$g'(x) = cxc^{-1}.$$

From that it follows that the mapping g can be written as

$$g(x) = \varepsilon cxc^{-1} \tag{4.54}$$

Making the following substitutions

$$\varepsilon c = a, c = b,$$

we can see that formula (4.51) is true, and therefore, the first part of Theorem 2 has been proven. ∎

Let us move on to the proof of the second part of the theorem.

Because the quaternion ε is defined by the mapping g in a unique way, then from the identity of two rotations g_1 and g_2 of the Euclidean vector space K^4 into itself follows the identity of the corresponding mappings g_1' and g_2'.

Therefore, due to formula (4.54), quaternions c_1 and c_2 that correspond to the mapping g_1 and g_2 according to Theorem 1 can be different only in sign.

From this it follows that mappings $f_{a,b}$ and $f_{a',b'}$ can be identical if and only if the quaternions a, b are identical to the quaternions a', b', or both of a, b differ from a', b' by the same sign.

So Theorem 2 has been proven. ∎

Let us show now that not every isomorphic mapping of the Euclidean vector space I onto itself is a rotation.

H) Quaternion units i, j, k from I constitute an orthonormal system that can be used as a basis of the space I. The $i, j, -k$ quaternions possess the same quality. Therefore, there exists an isomorphic mapping f of the three-dimensional Euclidean space I, such as

$$f(i) = i, \ f(j) = j, \ f(k) = -k.$$

The mapping f cannot be written as f_a (see (4.45)). Suppose it were the case, then we would have

$$f_a(i) = i, \ f_a(j) = j, \ f_a(k) = -k.$$

But it is impossible. Indeed,

$$f_a(k) = f_a(ij) = f_a(i)f_a(j) = ij = k.$$

So not every isomorphic mapping of the Euclidean vector space I onto itself can be represented as f_a, but due to Theorem 1 any rotation of the space I can be written in the form f_a. Thus not every isomorphic mapping f of the Euclidean vector space I is a rotation.

Let us prove now invariance of the definition of vector multiplication $[x, y]$ of two vectors x, y from I, that have been given in the coordinate form using an orthonormal basis i, j, k. (see Section 4.1, F)).

There is a basis

$$i, j, k \tag{4.55}$$

in the Euclidean vector space I, and also an orthonormal basis

$$u, v, w \qquad (4.56)$$

(see statement B) that can be derived from basis (4.55) by rotating it in the Euclidean vector space I (see statement E)). Also the rule of quaternion multiplication for (4.56) is identical with the rule of quaternion multiplication for (4.55).

Let now x and y be two quaternions from I. Let us write them in basis (4.55) and basis (4.56):

$$\begin{aligned} x &= x^1 i + x^2 j + x^3 k \\ y &= y^1 i + y^2 j + y^3 k. \end{aligned} \qquad (4.57)$$

Furthermore, we have

$$\begin{aligned} x &= \xi^1 u + \xi^2 v + \xi^3 w \\ y &= \eta^1 u + \eta^2 v + \eta^3 w \end{aligned} \qquad (4.58)$$

The vector multiplication $[x, y]$ using coordinates associated with basis (4.55) has been defined by the formula

$$[x, y] = (x^2 y^3 - x^3 y^2)i + (x^3 y^1 - x^1 y^3)j + (x^1 y^2 - x^2 y^1)k.$$

After that we have proven that quaternion multiplication xy may be represented by the formula

$$xy = -(x, y) + [x, y],$$

and the proof has been based on the rule of multiplication of quaternions (4.55).

Let us produce the quaternion multiplication of xy using basis (4.56).

Taking into account that the rules of multiplying quaternions (4.56) are identical with the rules of multiplying quaternions (4.55), then the quaternion result of xy will be written using coordinates (4.58) in the form

$$xy = -(x, y) + [x, y].$$

But instead of coordinates (4.57) there will be coordinates (4.58). The coordinate representation of the scalar product (x, y) is invariant relative to an orthonormal basis, so the vector product $[x, y]$ of two vectors x and y in both these orthonormal bases, that can be transformed into one another by some rotation, are also the same.

But this is correct only for the orthonormal bases that can be transformed one into another by the way of rotation.

CHAPTER 5

Other generalizations of numbers

In order to examine the question of number extensions rigorously in the first place we have to give the exact definition of those rules of operations on numbers that we consider intrinsic to numbers. These rules turn out to be the rules of operations on algebraic division rings and fields. Therefore, the first section of this chapter is devoted to a description of algebraic division rings and fields.

In the second section of the chapter the Frobenius Theorem is proven, which states that there are no other extensions of real numbers than complex numbers and quaternions.

5.1   Algebraic division rings and fields

In this book we have already dealt with several types of numbers, or let us say, elements, because quaternions are not usually called numbers. I mean rational, real, complex numbers, and quaternions. All these elements share some common properties. In each of these sets there are two operations, addition and multiplication. Each of these operations obeys some particular set of rules. Before formulating these rules, let us say that a set of elements on which are defined two operations, addition and multiplication, that comply with such rules is called a *division ring* in algebra, and if the operation of multiplication is commutative the set is called a *field*.

Let us denote K as a set of all elements that constitute a division ring. Let us now formulate the rules that the operations on K should comply with.

1. The addition operation on the division ring K is commutative. That means that if x and y are two elements of K, then the following identity holds

$$x + y = y + x. \tag{5.1}$$

2. The addition operation is associative. That means that if x, y, and z are of K, then the following identity takes place

$$(x + y) + z = x + (y + z).$$

3. In K there is a single element called the identity zero, and denoted as 0. This element satisfies the following condition:

$$x + 0 = x,$$

and therefore, (see 1.), $0 + x = x$.

4. For each element x of K there is an opposite element $-x$ that satisfies the following condition:

$$x + (-x) = 0.$$

From these rules it follows that the relationship

$$x + y = z \tag{5.2}$$

(where y and z are given elements), can be considered as an equation in which x is solvable. To solve for x let us add $-y$ to both parts of equation (5.2). We get then

$$x = z + (-y) = z - y.$$

Therefore, on K there is a subtraction operation that is the inverse operation to addition.

From all these we may see that the elements of K is a commutative group with respect to the addition operation.

For multiplication the same rules as for addition are valid with the exception of commutativity (the first one).

5. Multiplication in K is associative. In other words, if x, y, and z belong to K, then the following relationship is true

$$(xy)z = x(yz).$$

6. In K there is the identity element e such that

$$ez = ze = z.$$

7. For every element z of K that is not 0 there is a inverse element, that we denote as z^{-1}, such that

$$z^{-1}z = zz^{-1} = e.$$

From the rules of multiplication in K, it follows that relationships

$$xy = z, \tag{5.3}$$

$$yx = z, \tag{5.4}$$

where y and z are some given elements of K, and $y \neq 0$, may be solved for x. To solve equation (5.3) let us multiply both its parts from the right by y^{-1}. We get then

$$x = zy^{-1}.$$

To solve equation (5.3) let us multiply both its parts from the left by y^{-1}. We get then

$$x = y^{-1}z.$$

Therefore, with respect to multiplication the elements of K constitute a group.

And finally, the operations of addition and multiplication on K are bound by the following rule.

8. If x, y, z are three arbitrary elements of K, then the following relationships hold

$$(x + y)z = xz + yz;$$

$$z(x + y) = zx + zy.$$

From the rules it follows that

$$0z = z0 = 0. \qquad (5.5)$$

Indeed, we have

$$0z = (0 + 0)z = 0z + 0z.$$

Subtracting from the both parts of this relation $0z$ we get

$$0 = 0z.$$

The second relationship of (5.5) can be proved analogously. On the other hand, from the relationship

$$xy = 0 \qquad (5.6)$$

it follows that either x or y is equal to 0.

To formulate this property of a division ring, we say that in K there are no divisors of zero.

Let us suppose that $y \neq 0$, and let us prove then x is equal to 0. Multiplying both parts of equation (5.6) from the right by y^{-1}, we get $x = 0y^{-1} = 0$. We can prove in the same way that if $x \neq 0$ then $y = 0$.

Multiplication of an arbitrary element x of K by a whole non-negative number is naturally defined as follows,

$$0x = 0;$$
$$1x = x;$$
$$2x = x + x;$$
$$3x = x + x + x;$$

$$\ldots$$

It is easy to check that if m and n are two whole non-negative numbers, then

$$mx + nx = (m + n)x; \tag{5.7}$$

$$(mx)(ny) = (mn)(xy) = mnxy. \tag{5.8}$$

Let us define now the so-called characteristic of a division ring K that is equal either to 0 or to some prime number p. In order to do that let us make up a sequence of elements of the division ring K

$$0e, 1e, 2e, ..., ne, ... \tag{5.9}$$

If in sequence (5.9) only the single element $0e$ is equal to 0, then the division ring K is considered to have characteristic of 0.

If this is not the case, then in sequence (5.9) there are some elements equal to zero that are not $0e$. In sequence (5.9) the operations of addition and multiplication (7, 8) are defined. It turns out that when the characteristic of K is not equal to 0, sequence (5.9) contains a finite number of non-identical elements that form a field P^p, a residue class of a prime modulo p (see Section 5.2.)

In that case the characteristic of a division ring is that prime number p.

An example: it is clear that the field R of all rational numbers has its characteristic equal to 0. It turns out that in some sense the field R is the most primitive field of characteristic 0.

Definition 1 (Isomorphic division rings):
Two division rings K and $K^{'}$ are considered to be isomorphic if there exists a one-to-one mapping of K to $K^{'}$ that transforms the operations that are defined on K into the same operations on $K^{'}$. More specifically, there is such a one-to-one mapping f of the set K onto the set $K^{'}$ that preserves the results of the operations of addition and multiplication. In

other words, the following conditions are satisfied

$$f(x + y) = f(x) + f(y),$$

$$f(xy) = f(x)f(y).$$

The mapping f of K onto K' is called isomorphic, or an isomorphism.

It is clear that an isomorphism maps zero to zero, one-to-one, an opposite element to the opposite one, and a inverse element to the inverse one.

While defining the characteristic of division ring K we examined sequence (5.9) of elements of the division ring K. In the case that the characteristic of a division ring K is equal to zero, using the subtraction and division operations that are present in the division ring K, we can naturally define an operation of multiplication of any rational number r by the identity element e. Therefore, the division ring K includes all the elements of type re, where r is a rational number. The set of all such elements naturally constitute a field that is contained in the division ring K, and this field is isomorphic to the field of rational numbers. Thus the field of rational numbers is the most primitive division ring with characteristic equal to zero.

Definition 2 (subring and ring expansion):
If a division ring K includes a subset K_1, such that with respect to the operations of addition and multiplication defined on K it is also a division ring, then the division ring K_1 is called a subring of K, and K is called an expansion of the division ring K_1.

Let us now give a description of of the simplest field of characteristic p. It is the residue class modulo p.

5.2 The field of the residue class of a prime modulo p

A) Congruence of whole numbers.

Let $m > 1$ be some integer. We say that whole numbers a and b are congruent with each other modulo m and write

$$a \equiv b(mod\, m),$$

if the difference $b - a$ is divisible by m, in other words, if

$$b = a + cm, \qquad (5.10)$$

where c is a whole number.

It is clear that if two numbers are congruent with the same number modulo m, they are congruent with each other modulo m.

B) A set of all numbers congruent with some given number a modulo m is called a residue class of a modulo m. Let us denote it as $[a]$.

It is obvious that any two numbers of the set $[a]$ are congruent with each other modulo m. If b is some arbitrary number from the set $[a]$, then the sets $[a]$ and $[b]$ are identical.

$$[a] = [b].$$

It is clear that the set

$$[0], [1], [2], ...[m - 1]$$

represents a set of all residue classes modulo m. Therefore, there are exactly m residue classes modulo m.

C) Let a_1 and a_2 be two whole numbers, $[a_1]$ and $[a_2]$ are residue classes modulo m that include these numbers; b_1, and b_2 are two whole numbers that belong to the residue classes $[a_1]$ and $[a_2]$ respectively. Then according to (5.10) we have

$$b_1 = a_1 + c_1 m, b_2 = a_2 + c_2 m.$$

And therefore,

$$b_1 + b_2 = a_1 + a_2 + (c_1 + c_2)m.$$

From this we can see that the sum $b_1 + b_2$ belongs to the residue class $[a_1 + a_2]$. Therefore, adding any arbitrary number that belongs to the residue class $[a_1]$ to an any arbitrary number that belongs to the residue class $[a_2]$, we always get a number that belongs to the residue class $[a_1 + a_2]$.

Therefore, the residue class $[a_1 + a_2]$ is defined in a unique way by the residue classes $[a_1]$ and $[a_2]$. It does not depend on an arbitrarily chosen pair b_1 and b_2 of numbers from the residue classes $[a_1]$ and $[a_2]$. The residue class $[a_1 + a_2]$ is considered the sum of the residue classes $[a_1]$ and $[a_2]$:

$$[a_1] + [a_2] = [a_1 + a_2].$$

So in the set of all residue classes modulo m there is defined the operation of addition. The zero element of this addition is $[0]$. In exactly the same way we may define subtraction of two residue classes modulo m, and multiplication of two residue classes of modulo m. The residue class that represent one (the identity element) of this multiplication will be $[1]$.

We see that in the set of all residue classes modulo m there are defined operations of addition, subtraction, and multiplication.

D) If m is not a prime number, in other words, if it can be divided by some divisors

$$m = rs,$$

where r and s are both not equal to 1, then it is clear that

$$[r][s] = 0$$

Therefore, in case m is not a prime number, in the set of all residue classes of m there are "divisors of zero".

In that case, the set of all residue classes of m with the defined operations is not a field.

It turns out that if $m = p$ is a prime number, then the set of all residue classes of p is a field.

Let

$$[1], [2], ..., [p-1] \qquad (5.11)$$

be a set of all residue classes of p not equal to zero. The product of multiplication of two residue classes $[r]$ and $[s]$ from (5.11) can not be zero. If

$$[r][s] = [0],$$

then it would mean that rs can be divided by p and its residue class is equal to zero, and therefore, does not belong to sequence (5.11).

If $[a_1]$, $[a_2]$, and $[a]$ are three arbitrary residue classes of sequence (5.11), then the equality

$$[a_1][a] = [a_2][a] \qquad (5.12)$$

is possible only in the case when $[a_1] = [a_2]$. Indeed, from equality (5.12) it follows that

$$([a_1] - [a_2])[a] = [0].$$

Because $[a]$ is not equal to zero, then the other multiplier is equal to zero, and, thus $[a_1] = [a_2]$. Therefore, multiplying all the residue classes of sequence (5.11) by the same residue class $[a]$ we get exactly $p-1$ residue classes, and all of them will be different. That means that the multiplication of sequence (5.11) by $[a]$ maps it onto itself in a one-to-one way. Therefore, there is a residue class $[a']$, such that

$$[a][a'] = [1].$$

That means that for the residue class $[a]$ there is an inverse class $[a']$, and therefore, the set of all residue classes of prime modulo p is a field.

This field is called a field of residue classes modulo p, and p here is some prime number. We will denote this field as P^p.

E) If in the sequence (5.9) of elements of some division ring K there is some element equal to zero but different from $0e$, then denote m as the minimal integer such that $me = 0$. It turns out that if a whole number t satisfies the condition $te = 0$, then t can be divided by m.

Suppose that dividing t by m we get

$$t = qm + r,$$

where q is the quotient, r is the reminder, and $r < m$. Multiplying this equality by e from right, we get:

$$0 = 0 + re.$$

Therefore, $re = 0$, and $r < m$ which contradicts our assumption.

It is clear that if two whole numbers a and b have the property that $ae = be$, then a and b are congruent modulo m. Therefore, we can establish a one-to-one mapping between elements of a field K of type (5.9) from Section 5.1, and the residue classes of modulo m, that preserves the multiplication and addition operations. Because in the field K there are no divisors of zero, then the number m should be a prime number (see D)). Therefore, the set of elements of type (5.9) of the field K is isomorphic to the field of residue classes P^p of a prime number p.

The prime number p is called the characteristic of the field K. And therefore, the field P^p is the simplest field of characteristic p.

Statement D) is directly connected to the so called Fermat's Small Theorem, which is different from the famed Fermat's Great Theorem. Fermat's Small Theorem will not be used is this book. I give it here along with the proof though, as an example of a non-trivial application of residue classes.

Theorem 3 (Fermat's Small Theorem):
Let p be a prime number and a be an arbitrary whole number, not divisible by p. Then the following relation holds:

$$a^{p-1} \equiv 1 (mod\, p). \qquad (5.13)$$

Sequence (5.9) comprises all elements of P^p that are not zero. Multiplying all these elements by a residue class $[a]$ from P^p, as it has been proven in the statement D), we get the same elements of sequence (5.9) but in some other order. Therefore, in the field P^p there is the following equality:

$$[a]^{p-1}[1][2]...[p-1] = [1][2]...[p-1].$$

The result in the right side of the equality is not a zero of the field P^p, so we can divide both parts of the equality by it. In this way we get the equivalent equality

$$[a]^{p-1} = [1].$$

This equality that takes place in the field P^p is in turn an equivalent of the number equality (5.13) that represents the statement of Fermat's Small Theorem. Therefore, Fermat's Small Theorem has been proven. ∎

Example 1:
Let us check Fermat's Small Theorem on some simple numerical example. Let $p = 5$, $a = 2$. We have $2^4 = 16$, and 16 is congruent with 1 modulo 5.
Let us consider now simple residue classes modulo 2, 3, and 4. There are two residue classes modulo 2: [0] and [1]. For these residue classes we have the following rules for the addition and multiplication operations:

$$[0] + [1] = [1], [1] + [1] = [0],$$

$$[0][1] = [0], [1][1] = [1].$$

From these rules we can see that the residue classes modulo 2 represent a field that we denote as P^2.

There are three elements in the residue class modulo 3: $[0]$, $[1]$, $[2]$. *Let us write down the rules of multiplication and addition for these residue classes. We have*

$$[1] + [1] = [2], [1] + [2] = [0], [2] + [2] = [1],$$

$$[1][1] = [1], [2][2] = [1].$$

From these rules we can see that the residue classes modulo 3 represent a field, that we denote as P^3.
There are four residue classes modulo 4: $[0]$, $[1]$, $[2]$, $[3]$.
Let us write down the rules of multiplication and addition for these residue classes. We have

$$[1] + [2] = [3], [1] + [3] = [0],$$

$$[2] + [3] = [1], [3] + [3] = [2],$$

$$[2][2] = [0], [2][3] = [2], [3][3] = [1].$$

From these rules we can see that the residue classes modulo 4 do not represent a field, because the square of the residue class $[2]$ *is equal to zero. Specifically, this is the reason why we cannot divide both parts of the equality* $[2][3] = [2]$ *by* $[2]$, *because for the element* $[2]$ *there is no inverse element.*

5.3   Frobenius Theorem

Theorem 4 (Frobenius Theorem):
Let L *be a division ring that includes as a subring the division ring* D *of real numbers, and for each element of* L, *multiplication by an element of* D *is commutative, and also each element* x *of* L *can be written as*

$$x = x^0 + x^1 i_1 + ... + x^n i_n, \qquad (5.14)$$

where $x^0, x^1, ..., x^n$ *are the real numbers that are coordinates of* x, *so* x *is an* $(n+1)$ *dimensional vector. Therefore, it is assumed that*

$$1, i_1, ..., i_n$$

constitute a basis of the vector space L. In order to define multiplication in L, it is enough to set up a rule of multiplication of $i_1, ..., i_n$ in such a way that every result $i_r i_s$ can be written in form (5.14). If these conditions are met it turns out that L is either identical with the field D, in other words it is isomorphic to the field of real numbers, or it is isomorphic to the K^2 field of complex numbers, or it is isomorphic to the K^4 division ring of quaternions.

The field D of real numbers comprises elements of type (5.14) for which only coördinate x^0 can be different from zero. The set D is a one dimensional vector subspace of the vector space L. The characteristic of the division ring L is obviously 0.

In order to make the proof more manageable let us split it into sections A), B), and C).

A) In the division ring L let us distinguish some elements that we will consider as pure imaginary. Let us denote the set of all such elements as I. To L we assign all such z of L such that the square of z is a real number, or in other words it belongs to D, and is non-positive. Therefore, I comprises all such z that

$$z^2 \in D; z^2 \leq 0$$

The last inequality can change to an equality if and only if $z = 0$. It is clear that the intersection of I and D is zero.

The elements of I have the following properties.

If a is a real number, and $z \in I$, then

$$az \in I. \tag{5.15}$$

Further, if $z \in I$, and $z \neq 0$, then

$$z^{-1} \in I. \tag{5.16}$$

It turns out that each element x of L can be written in a single way as a sum

$$x = a + z, \tag{5.17}$$

where $a \in D$, $z \in I$.

Let us prove statement A). Let us start from the statement (5.15). We have $(az)^2 = a^2 z^2$. Because $a^2 \geq 0, z^2 \leq 0$ then $(az)^2 \leq 0$, and therefore, $az \in I$.

To prove (5.16) let us consider the multiplication

$$z^2(z^{-1})^2 = zzz^{-1}z^{-1} = 1.$$

Thus $(z^{-1})^2 = (zz)^{-1}$, and therefore, is negative. So $z^{-1} \in I$.

Let us now prove statement (5.17). Let us make up a sequence

$$1, x, x^2, ..., x^{n+1}. \tag{5.18}$$

Since the dimension of the vector space L is equal to $n+1$, but the number of elements of (5.18) is $n+2$, then according to Statement A) of Section 4.1 the elements of this sequence are linearly dependent, in other words the following relationship holds

$$\alpha_0 + \alpha_1 x + ... + \alpha_{n+1} x^{n+1} = 0,$$

where coefficients $\alpha_0, \alpha_1, ..., \alpha_{n+1}$ are real numbers and not all of them are equal to zero. Therefore, the element x is a root of a polynomial with real coefficients, and according to the results of Section 3.2 is either a root of a polynomial g_1 of first degree, or a polynomial g_2 of second degree. So for x one of the following conditions is true

$$x - \alpha = 0; \ x^2 - 2ax + b = 0,$$

where $a^2 - b < 0$. In the first case x is equal to the real number α, and for decomposition (5.17) we have $a = \alpha, z = 0$. For the second case we have $(x - a)^2 = a^2 - b < 0$. Therefore, the element $x - a$ belongs to I, let us denote is as z. We get, therefore, $x = a + z$, and decomposition (5.17) is proven. ■

Let us prove now that decomposition (5.17) is unique. Suppose that along with (5.17) we also have a decomposition

$$x = a_0 + z_0, \tag{5.19}$$

where $a_0 \in D, z_0 \in I$. Let us prove in that case that $a_0 = a$, and $z_0 = z$. Subtracting (5.19) from (5.17) we get

$$z = z_0 + (a_0 - a).$$

Taking the square of each side of this equality we get

$$z^2 = z_0^2 + 2z_0(a_0 - a) + (a_0 - a)^2,$$

or

$$2z_0(a_0 - a) = z^2 - z_0^2 - (a_0 - a)^2.$$

Here on the left side is a pure imaginary value, but on the right side is real, therefore, both of them are equal to zero. Thus we have $2z_0(a_0 - a) = 0$. Because the characteristic of the division ring L is not equal to 2, then $z_0(a_0 - a) = 0$ which is possible only if one of the multipliers is equal to zero. First, suppose that $z_0 = 0$. Then from relationship (5.19) it follows that x is the real number a_0. From relationship (5.17) we get $z = a_0 - a$. Thus the left part is a pure imaginary value, and the right side is a real one. Therefore, both of them are equal to zero. So we have come to the conclusion that $z_0 = z = 0$, and from this follows that $a_0 = a$. Second, if the multiplier $a_0 - a$ is equal to zero, then $a_0 = a$, and from this according to (5.17) and (5.19) it follows that $z_0 = z$. Therefore, we have proven the uniqueness of decomposition (5.17), and statement A) is fully proven. ■

B) Let u and v be two elements of I, and let r and s be two real numbers. Then the following two relationships hold:

$$\xi = uv + vu \in D, \tag{5.20}$$

$$\eta = ru + sv \in I. \tag{5.21}$$

From relation (5.21) it follows that I is a vector subspace of the vector space L. From this and from statement A) it follows that L decomposes into the direct product of its subspaces D and I.

Let us prove statement B). Let us consider first a simple case when elements $u, v, 1$ are linearly dependent. In that case we have

$$\alpha u + \beta v + \gamma 1 = 0,$$

and not all of the real numbers α, β, γ are equal to zero. Let us rewrite the last equality as

$$\alpha u = -\beta v - \gamma.$$

Here αu and $-\beta v$ are pure imaginary elements according to relation (5.15), and because of the uniqueness of decomposition (5.17) we get $\gamma = 0$. Therefore,

$$\alpha u = -\beta v.$$

Since both of the coefficients α and β can not be zero, at least one of them is not zero. Suppose $\beta \neq 0$. Then

$$v = -\frac{\alpha}{\beta} u.$$

Putting this formula in (5.20), we get

$$\xi = -\frac{2\alpha}{\beta} u^2,$$

so ξ is a real number, and relation (5.20) has been proven for this case. Putting the derived expression for v in (5.21) we get

$$\eta = (r - \tfrac{s\alpha}{\beta})u.$$

So taking into account (5.15) we can see that η is a pure imaginary element.

Let us now consider the case when $u, v, 1$ are linearly independent. We will prove statement (5.20) together with (5.21). Let us consider only the case when both r and s are not equal to zero. In the opposite case both statements (5.20) and (5.21) can be easily checked. Let us decompose elements ξ and η according to (5.17). Let

$$\xi = uv + vu = a + z, \tag{5.22}$$

where $a \in D, z \in I$;

$$\eta = ru + sv = a_0 + z_0, \tag{5.23}$$

where $a_0 \in D, z_0 \in I$.

In order to prove relations (5.20) and (5.21) it is enough to prove that $z = 0$ and $a_0 = 0$.

Let us square both sides of (5.23). We get

$$r^2u^2 + s^2v^2 + rs(uv + vu) = a_0^2 + 2a_0z_0 + z_0^2.$$

In the left side of this relation, replacing $uv + vu$ by its expression according to equality (5.22), we get

$$r^2u^2 + s^2v^2 + rs(a + z) = a_0^2 + 2a_0z_0 + z_0^2. \tag{5.24}$$

The left and the right sides of this equality are combinations of real and pure imaginary numbers. Because of the singularity of decomposition (5.17) both these parts should be equal respectively to the parts on the other side. The pure imaginary part on the left side is rsz, and the pure imaginary part on the right side is $2a_0z_0$, so we have the equality

$$rsz = 2a_0z_0. \tag{5.25}$$

Let us now consider two distinct cases. First, $z = 0$; second, $z \neq 0$.

In the first case, from equality (5.25) we have $a_0z_0 = 0$. Therefore, either a_0 or z_0 is equal to zero. If $a_0 = 0$, then relation (5.21) is true, and relation (5.20) follows from the assumption that $z = 0$. If $z_0 = 0$ then equality (5.23) becomes a linear dependency between u, v and 1. This case we have already investigated. So in case $z = 0$ we have proven statement B).

Let us now consider the case when $z \neq 0$. Because both r and s are not equal to zero, then from relation (5.25) it follows that $a_0 \neq 0$, and relation (5.25) can be rewritten as

$$z_0 = \frac{rs}{2a_0}z.$$

Putting this expression for z_0 into equality (5.23) we get

$$ru + sv = a_0 + \frac{rs}{2a_0}z. \tag{5.26}$$

Note that z does not depend on numbers r and s, and equality (5.26) has been proven in the case that $z \neq 0$. Let us write now equality (5.26) for some other numbers r' and s'. We get

$$r'u + s'v = a_0' + \frac{r's'}{2a_0'}z \tag{5.27}$$

Let us remove z from the equalities (5.26) and (5.27), by multiplying equality (5.27) by some proper real number c and then subtracting it from equality (5.26).

We get

$$(r - cr')u + (s - cs')v = a_0 - ca_0'. \tag{5.28}$$

Because numbers r' and s' were chosen randomly only on the condition that both of them are not zero, we can choose them in such a way that they are not proportional to each other. In that case, in equality (5.28) the coefficients $r - cr'$ and $s - cs'$ cannot both be zero. Then equality (5.28) becomes a linear dependency between elements u, v and 1, in other words it turns into the case we already proved in the very beginning of the proof of statement B).

Therefore, statement B) has been proven. ■

C) Let u and v be two elements of I such that

$$u^2 = -1, \ v^2 = -1, \ w = uv \in I. \tag{5.29}$$

In that case u, v, w satisfy the same criteria as quaternion identity elements (quaternion ones, see (4.34), (4.35)), in other words, the following equalities hold

$$u^2 = v^2 = w^2 = -1, \tag{5.30}$$

$$\begin{aligned} uv &= -vu = w, \\ vw &= -wv = u, \\ wu &= -uw = v. \end{aligned} \tag{5.31}$$

Let us prove statement C). First of all let us note that

$$vu = (uv)^{-1}$$

Indeed,

$$uvvu = u(-1)u = u^2(-1) = 1.$$

Therefore, we have

$$uv + vu = uv + (uv)^{-1}. \tag{5.32}$$

Taking into account (5.20) we see that the left side of this equality is a real number. The right side is a sum of two pure imaginary elements, because $(uv)^{-1}$, due to (5.16), is a pure imaginary element, and as such, due to 5.21, is a pure imaginary one. Therefore, both sides of the equality (5.32) are zeros, and thus we have:

$$uv = -vu. \tag{5.33}$$

So we have proven the first of equalities (5.31). Let us prove now the last of equalities (5.31). We have

$$w^2 = uvuv = -uvvu = -1.$$

Let us prove now the second and the third equations of relationship (5.31). We have

$$vw = vuv = -vvu = u. \tag{5.34}$$

Three elements v, w, and $u = vw$ satisfy the same conditions as the elements of (5.29), and for the elements of (5.29), relation (5.33) holds. Therefore, we have

$$wv = -vw.$$

Thus this relationship, combined with (5.34), gives us the second equation of relationship (5.31).

Let us now consider

$$wu = uvu = -uuv = v.$$

Exactly as we did it before we can prove that

$$wu = -uw.$$

Thus, the third equation of the relationships (5.31) has been proven.

Therefore, statement C) has been proven. ∎

Let us now move to the final stage of the proof of Theorem 4.

If I contains only zero, then L is identical to D, and therefore, is isomorphic with the field of real numbers D. Thus in that case Theorem 4 is true.

Now suppose that in I there is a vector z_0 not equal to zero. Then denote

$$i = \frac{z_0}{\sqrt{-z_0^2}},$$

we get the element $i \in I$ such that

$$i^2 = -1.$$

If the dimension of I is equal to 1, then any element of I can be written as

$$bi$$

where b is a real number. In that case any element of L can be written as

$$a + bi.$$

Therefore, in the case where the dimension of the vector space I is equal to 1, then the division ring L is isomorphic to the field of complex numbers K^2.

Now suppose that the dimension of I is bigger than one. In that case there exists an element z_1 in I, such that the pair of vectors

$$i, z_1$$

is linearly independent.

Let us consider the product iz_1. According to (5.17)

$$iz_1 = a + z.$$

Let $j_1 = z_1 + ai$. Then we have

$$ij_1 = iz_1 - a = z.$$

101

Denoting j as

$$j = \frac{j_1}{\sqrt{-j_1^2}},$$

we get for j the following equality

$$j^2 = -1.$$

Therefore, elements i and j satisfy the following conditions

$$i^2 = -1, \; j^2 = -1, \; i, \, j \in I.$$

Thus, elements

$$i, \; j, \; k = ij$$

according to statement C satisfy the same conditions as the elements u, v, w. Therefore, a set of all elements that can be written as

$$x = x^0 + x^1 i + x^2 j + x^3 k,$$

where x^0, x^1, x^3 are real numbers, constitutes the division ring K^4 of quaternions.

Let us now suppose that the dimension of the space I is bigger than three, and let us show that this assumption leads to a contradiction.

If the dimension of the space I is bigger that three, then there exists an element z_2, such that vectors

$$i, \; j, \; k, \; z_2$$

are linearly independent. According to decomposition (5.17) we have

$$iz_2 = a + x, \; jz_2 = b + y, \; kz_2 = c + z,$$

where a, b, c are real numbers, and x, y, z are belong to I. Let us make up the element

$$l_1 = z_2 + ai + bj + ck.$$

According to (5.21) $l_1 \in I$, and due to linear independence of vectors i, j, k, z_2 we have $l_1 \neq 0$.

Besides that, we have

$$il_1 = iz_2 - a + bk + cj = x + bk + cj \in I$$

according to (5.21). Exactly in the same way it can be proven that $jl_1 \in I$ and $kl_1 \in I$.

Denoting l as

$$l = \frac{l_1}{\sqrt{-l_1^2}},$$

then

$$l^2 = -1,$$

and elements $i, l, il \in I$ satisfy the same conditions as the elements u, v, w from statement C). So we have

$$il = -li.$$

In the same way we can prove that

$$jl = -lj \qquad \text{and} \qquad kl = -lk.$$

Let us consider now the element ilj. We have

$$ilj = i(-jl) = -kl, \; ilj = -lij = -lk = kl.$$

Therefore, we come to the conclusion that $2kl = 0$, but this is impossible because the characteristic of the division ring L is not equal to two.

Therefore, Theorem 4 has been fully proven. ∎

CHAPTER 6

Topologo-algebraic division rings

The division rings we have investigated so far, the field of real numbers, the field of complex numbers, and the division ring of quaternions are Euclidean vector spaces of dimensions 1, 2, and 4 respectively. Because there is a metric in these vector spaces, in other words there is a defined distance between any two points (see chapter 4), then there is also a notion of convergence. Let us denote any of the above mentioned division rings as K. If

$$a_1, a_2, ..., a_n, ...$$

is some sequence of elements of K, then we know what it means that this sequence converges to an element a. It means that the distance $\varrho(a_n, a)$ between points a_n and a tends to zero with increasing n. This can be written as a formula in this way:

$$\lim_{n \to \infty} a_n = a. \tag{6.1}$$

We will express the fact that there is a convergence in the space K by saying that K is a topological space.

It is easy to prove that the algebraic operations that exist in the division ring K are continuous with respect to the topology that exists there. More precisely, if the sequence

$$b_1, b_2, ..., b_n, ...$$

of elements of the division ring K converge to an element b, in other words, if

$$\lim_{n \to \infty} b_n = b,$$

then the conditions for continuity of the operations of addition and multiplication are met. In particular, the following relations hold:

$$\lim_{n \to \infty} (a_n + b_n) = (a + b), \quad \lim_{n \to \infty} a_n b_n = ab.$$

This shows that the addition and multiplication operations that exist in K are continuous with respect to existing in the K topology.

Because the operations of subtraction and division can be reduced to finding the opposite and the inverse element to a given element, the continuity of the subtraction and division operations may be formulated in this way:

$$\lim_{n \to \infty} (-a_n) = -a.$$

If $a \neq 0$, then

$$\lim_{n \to \infty} a_n^{-1} = a^{-1}.$$

If in some algebraic division ring L there is a topology, or in other words, there exists convergence, or putting it in other words, there is a defined meaning for relation (6.1), and operations that exist in the division ring L are continuous with regard to this topology, then such a division ring is called topological algebraic, or in short, a topological division ring if it is already known that it is an algebraic division ring.

The study of topological division rings is a part of topological algebra. Of interest about topological division rings is that there exist only a few of them and all of them can be reasonably described. Section (6.1) will give an accurate definition of the topological division ring.

It is clear that relation (6.1) will occur if all elements of the sequence

$$a_1, a_2, ..., a_n, ...,$$

are identical to each other starting at some index. But if relation (6.1) occurs only in this condition, then the existence of convergence does not introduce anything new in the algebraic division ring L, so we are not going to examine such division rings and will not call them topological.

6.1 Topological division ring

As was said at the beginning of this chapter, an algebraic division ring becomes topological if along with algebraic operations there is the operation of convergence, and if the algebraic operations are continuous with regard to convergence. The three kinds of division rings we have investigated so far are all Euclidean spaces, and the operation of convergence is defined for them using the notion of distance. In order to define that operation in Euclidean space, we have to know how to obtain the distance between any two points in the space. It is possible however to define the operation of convergence without using the notion of distance, which is the common approach for general topological division rings, but instead using the operation of subtraction that is defined in K. Indeed, the statement that the distance between a_n and a tends to zero is equivalent to the statement that $|a_n - a|$ is decreasing to 0. In order to define the operation of convergence (see (6.1)), let us denote as U_n the set of all elements x of the division ring K such that

$$|x| < \frac{1}{n}. \tag{6.2}$$

So U_n is an open ball with radius of $1/n$ with center at the origin. In order to comply with (6.1) it is necessary and sufficient to satisfy the following condition: for every natural number n there exists a natural number r such that for any $p > r$ we have

$$(a_p - a) \in U_n.$$

Let us extend this construction to any division ring L. But first let us introduce some new definitions.

A) If X and Y are two sets that belong to L then let us denote $X + Y$ as a set of all elements

$$x + y, \qquad \text{where} \qquad x \in X, \; y \in Y,$$

and X - Y as a set of elements

$$x - y, \qquad \text{where} \qquad x \in X, \; y \in Y.$$

Furthermore, denote XY as a set of elements

$$xy, \qquad \text{where} \qquad x \in X, \; y \in Y,$$

and XY^{-1} as a set of elements

$$xy^{-1}, \qquad \text{where} \qquad x \in X, \; y \in Y.$$

For $y = 0$ the element y^{-1} is not defined in the last formula, so we consider only those elements $y \in Y$ that are not equal to zero.

Definition 3:

An infinite decreasing sequence of sets

$$U_1, U_2, ..., U_n, ... \tag{6.3}$$

of L such that each set contains the zero of the division ring L,

$$0 \in U_{n+1} \subset U_n,$$

is said to be a complete neighborhood system in the topological division ring L if the following five conditions are satisfied.
a) For every natural number n there is a large enough number p such that

$$(U_p + U_p) \subset U_n.$$

b) *For every natural number n there is a large enough number p such that*

$$(U_p U_p) \subset U_n.$$

c) *For every natural number n there is a large enough number p such that*

$$(-U_p) \subset U_n.$$

d) *(Let us recollect that we denoted e as an identity element of the division ring L.) For every natural number n there is a large enough number p such that*

$$(e + U_p)^{-1} \subset e + U_n.$$

e) *For any element $a \in L$, and for every natural number n there exists a large enough number p such that*

$$U_p a \subset U_n, \ a U_p \subset U_n.$$

B) The operation of convergence in the division ring L can be defined, using the complete neighborhood system of zero (see (6.3)), in the same way as was done for the topological space K.

Definition 4:
A sequence

$$a_1, a_2, ..., a_n, ...$$

of elements of L is converging to to an element a,

$$\lim_{n \to \infty} a_n = a,$$

if for every natural number n there exists a large number r such that, for $p > r$, we have

$$(a_p - a) \in U_n.$$

Then the following theorem holds:

Theorem 5:

If convergence in L is defined in the way specified in B) then algebraic operations that exist in the division ring L are continuous with respect to this convergence. In particular, if the following relations are true

$$\lim_{n \to \infty} a_n = a, \ \lim_{n \to \infty} b_n = b \qquad (6.4)$$

then

$$\lim_{n \to \infty} (a_n + b_n) = a + b, \qquad (6.5)$$

$$\lim_{n \to \infty} (a_n b_n) = ab, \qquad (6.6)$$

$$\lim_{n \to \infty} (-a_n) = -a. \qquad (6.7)$$

If $a \neq 0$, then the following relation holds:

$$\lim_{n \to \infty} a_n^{-1} = a^{-1}. \qquad (6.8)$$

Therefore sequence (6.3) defines a topology in the division ring L.

Let us prove relation (6.5). From formula (6.4) it follows that for any large number r there exists a large number p such that if $p > r$ we have

$$a_p \in a + U_r, \ b_p \in b + U_r.$$

Then

$$a_p + b_p \in a + b + U_r + U_r. \qquad (6.9)$$

Due to condition a) of definition 3, for each integer n there is a natural number r large enough that

$$U_r + U_r \subset U_n.$$

Then, from (6.9) it follows that

$$(a_p + b_p) - (a + b) \in U_n.$$

Therefore relation (6.5) has been proved. ∎

Let us prove relation (6.6). From (6.4) it follows that for any natural number r there is a natural number p large enough that when $p > r$ we have

$$a_p \in a + U_r, \; b_p \in b + U_r.$$

Then

$$a_p b_p \in (a + U_r)(b + U_r) = ab + aU_r + U_r b + U_r U_r.$$

Because of conditions a), b), and e) of definition 3, for any natural number n there exists a natural number r so big that

$$aU_r + U_r b + U_r U_r \subset U_n.$$

Therefore

$$(a_p b_p - ab) \in U_n,$$

and, thus relation (6.6) has been proved. ∎

Relation (6.7) follows from condition c) of definition 3.

Let us prove relation (6.8). We will consider element $a_p^{-1} a$, but let us first investigate the reciprocal element $a^{-1} a_p$. From formula (6.4) it follows that for an arbitrary natural number q there is a natural number p large enough that

$$a_p \in a + U_q.$$

Further, for an arbitrarily large natural number s there is a natural number q large enough that

$$a^{-1}(a + U_q) = (e + a^{-1}U_q) \subset e + U_s$$

(see condition e) of definition 3). From this it follows that for a large enough natural number p the following relation holds

$$a^{-1}a_p \in (e + U_s).$$

Because of condition d) of relation 3, it follows that

$$a_p^{-1}a \in e + U_r,$$

where r is an arbitrary natural number, and p is a large enough natural number. Furthermore, we have

$$(a_p^{-1}a - e) \in U_r, \quad (a_p^{-1} - a^{-1}) \in U_r a^{-1}.$$

Because of condition e) of definition 3, for any natural number n there is a natural number r large enough that

$$U_r a^{-1} \subset U_n.$$

Therefore,

$$(a_p^{-1} - a^{-1}) \in U_n.$$

Therefore, relation (6.8) has been proved. Thus theorem 5 has been proved. ■

C) It turns out that the neighborhood system of zero

$$U_1, U_2, ..., U_n, ..., \tag{6.10}$$

which is defined by relation (6.2) on any of the three division rings of K we examined, satisfies, all conditions of definition 3.

Let us prove this statement. For the neighborhood system (6.10), we have

$$U_p + U_p \in U_n,$$

where $p > 2n$. Condition a) follows from this fact. Furthermore, we have

$$U_p U_p \subset U_{p^2}.$$

From this follows condition b) of definition 3. Furthermore, we have

$$-U_n = U_n.$$

From that we get condition c) of definition 3.

Furthermore, if a is an arbitrary element of the division ring K and $x \in U_p$, then

$$|ax| = |a| \cdot |x| < |a|\frac{1}{p}.$$

Therefore, for p larger then $n|a|$, condition e) of definition 3 is satisfied.

Let x be an arbitrary element of U_n (see (6.2)). Then the set

$$(e + U_n)^{-1}$$

is composed of elements of the following type

$$(e + x)^{-1}.$$

But we have

$$(e + x)^{-1} = e - x + x^2 - x^3 + ... \tag{6.11}$$

This formula is known to be true for the real and complex numbers, and it is also true for quaternions because the quaternions included in this formula are commutative in multiplication. Formula (6.11) may be rewritten as

$$(e + x)^{-1} = e + y,$$

where

$$y = -x + x^2 - x^3 + ...,$$

so

$$|y| \leq \frac{1}{n} + \frac{1}{n^2} + \frac{1}{n^3} + ... = \frac{1}{n-1}.$$

Therefore,

$$(e + U_n^{-1}) \in e + U_{n-1}.$$

From this follows condition d) of definition 3.

6.2 Topological concepts of topological division ring L

In this section will be introduced some topological concepts well known in topology. In this book I do not assume though the reader's acquaintance with these concepts.

A) Let M be some sub-set of the elements that constitute the space L. We will call a point $a \in L$ a limit point for the set M if in M there is a sequence

$$a_1, a_2, ..., a_n, ...$$

of not identical elements that converges to the element a.

The set M is called closed if all its limits belong to it.

Uniting an arbitrary set M with all its limits we get the closure of the set M that is denoted as $\bar{M}$. If a set M does not have any limit points, $\bar{M}$ consists of the set itself.

A set M is called closed if $M = \bar{M}$. It turns out that the closure of any set M is also closed, or

$$\bar{\bar{M}} = \bar{M}. \tag{6.12}$$

A proof of this relation is simple, but laborious. So I give here just some indication of how to prove it.

Let

$$b_1, b_2, ..., b_n, ... \qquad (6.13)$$

be a sequence of distinct points of $\bar{M}$ that converges to some point b. What we need to prove is that $b \in \bar{M}$. Without reducing the scope of the proof we may assume that all points of sequence (6.13) do not belong to M. For each point b_n there is a sequence B_n of distinct points of M that converges to it.

In the sequence B_n let us choose some point with a large enough index, and let us denote it as c_n. We get then a sequence

$$c_1, c_2, ..., c_n, ...$$

of distinct points of M that converges to b. Therefore b is a limit point of M, and therefore relation (6.12) has been proved.

■

In this way we can prove that the set $\bar{M}$ is closed.

B) A set Q of points of the space L is called compact if every infinite subset M of the set Q has at least one limit point that belongs to Q.

C) A topological division ring L is called locally compact if there exists a set U_n that belongs to the neighborhood system of zero (6.3), such that its closure $\bar{U}_n$ is compact.

D) A closed infinite set M of elements of L is called connected if it cannot be presented as a union of two non-empty subsets M_1 and M_2 that do not have any common elements.

That means that M cannot be split into a sum of two closed non-intersecting subsets M_1 and M_2.

E) It turns out that all three division rings K that we have investigated so far, i.e. the field of the real numbers, the field of the complex numbers, and the division ring of quaternions, are locally compact and connected.

Because all three division rings K are Euclidean spaces, we start the proof of statement E) with examination of an n-dimensional Euclidean space A^n that is composed of vectors

$$x = (\vec{x^1}, \vec{x^2}, ..., \vec{x^n}).$$

We accept without proof that the set of real numbers

$$-1 \leq t \leq 1$$

is compact and connected.

In the space A^n let us investigate the cube Q that is defined by the following conditions

$$-1 \leq x^i \leq 1, \; i = 1, 2, ..., n,$$

and let us prove that this cube is a compact set. In the cube Q let us consider an infinite sequence of mutually different points

$$x_1, x_2, ..., x_p, ... \tag{6.14}$$

Let

$$x_1^i, x_2^i, ..., x_p^i, ..., \; i = 1, ..., n \tag{6.15}$$

be a sequence of i coordinates of points (6.14). Note, that all numbers in this sequence belong to the interval $-1 \leq x \leq 1$, which is compact. Therefore from the sequence of numbers (6.15) for $i = 1$ it is possible to choose a convergent subsequence X^1, that, generally speaking, is not necessary a sequence of mutually different numbers.

The sequence X^1 corresponds to the subsequence Y^1 of vectors (6.14). The second coordinates of the sequence Y^1 is a sequence of numbers that belong to the interval $-1 \leq t \leq 1$. From this sequence of the second coordinates let us choose a convergent subsequence that corresponds to the subsequence Y^2 of vectors (6.14). Proceeding in this way, we get a subsequence Y^n of the vector sequence (6.14) that converges to some point y. Therefore, we have established that from the sequence of mutually different vectors (6.14) it is possible to choose a subsequence Y^n that converges to some point y that belongs to the cube Q. From this fact it immediately follows that the cube Q is compact. Indeed, if M is some infinite set of points of cube Q, then from M we can select a sequence (6.14) of mutually different vectors. This sequence converges to some vector y, which is a limit point of the set M, which means that the cube Q is compact.

In each of these three Euclidean spaces K that correspond to the three examined division rings K, the cube Q is a compact set. Any neighborhood set of type U_p (see (6.2)) belongs to the cube Q, and therefore, its closure $\bar{U}_p$ is compact. Therefore, all three investigated division rings K are locally compact.

Let us prove now that the Euclidean space A^n is connected. Suppose it is not true, in other words, that it can be split into two sets M_1 and M_2 that have no intersection. Let us consider a line segment in the space A^n

$$x(t) = (\frac{1}{2} - \frac{1}{2}t)x_{-1} + (\frac{1}{2} + \frac{1}{2}t)x_1,$$

$$-1 \leq t \leq 1,$$

(6.16)

such that the initial point of the segment $x(-1) = x_{-1}$ belongs to M_1, and the end point of the segment $x(1) = x_1$ belongs to M_2. Let us denote the intersection of this segment (6.16) with M_1 as M_1', and the intersection of this segment with M_2 as M_2'. The sets M_1' and M_2' are closed subsets of the line segment (6.16). Considering that A^n is split into a sum of two closed sets M_1 and M_2 which do not intersect, then the line segment is also split into a sum of two closed sets M_1' and M_2' that also do not intersect, but this contradicts our supposition that a line segment is a connected set.

So we have proved that any Euclidean vector space A^n is connected. Therefore all three division rings K that we have investigated are connected.

The statement E) has been proved. ∎

F) A sequence

$$a_1, a_2, ..., a_n, ...$$

(6.17)

of elements of the topological division ring L is called a

Cauchy sequence if for any natural number n there is a large enough natural number r, such that for $p > r$ and $q > r$, we have

$$(a_p - a_q) \in U_n.$$

It turns out that if sequence (6.17) converges in the topological division ring L, then it is a Cauchy sequence. However not every Cauchy sequence converges. But if some infinite subsequence of the Cauchy sequence converges then the Cauchy sequence itself also converges.

Let us prove that if sequence (6.17) converges then it is a Cauchy sequence.

If

$$\lim_{n \to \infty} a_n = a,$$

then from this it follows that for any natural number s there is a natural number r, such that for $p > r$ and $q > r$, we have

$$(a_p - a) \in U_s, \ (a_q - a) \in U_s$$

(see B) 6.1.) From this it follows that

$$a_p - a_q = (a_p - a) - (a_q - a) \in U_s - U_s.$$

Then due to conditions a) and c) of definition 3 for a large enough s we have

$$(U_s - U_s) \subset U_n,$$

where n is an arbitrary large natural number.

Therefore we have proved that sequence (6.17) that converges to the element a is a Cauchy sequence.

Let us prove now that if sequence (6.17) is a Cauchy sequence, and some of its subsequence converges to the element a, then sequence (6.17) itself converges to that element.

Because some infinite subsequence of sequence (6.17) converges to a, then in the sequence (6.17) there are elements with large numbers p such that

$$(a_p - a) \in U_s,$$

where s is some arbitrary large natural number. Taking into account that sequence (6.17) is a Cauchy sequence, then for the natural number s there exists a large enough natural number r, that for $p > r$ and $q > r$ we have

$$(a_p - a_q) \in U_s.$$

Because

$$(a_p - a) \in U_s,$$

then

$$a_q - a = (a_q - a_p) + (a_p - a) \in -U_s + U_s.$$

Therefore

$$(a_q - a) \in U_n,$$

and sequence (6.17) converges to a.

G) A topological division ring L is called complete if every Cauchy sequence there converges.

It turns out that every locally compact topological division ring L is complete.

Let

$$a_1, a_2, ..., a_p, ... \tag{6.18}$$

be some Cauchy sequence of the division ring L. Then according to the definition of Cauchy sequence (see F)) the difference $a_p - a_q$ for large enough p and q belongs to some neighborhood set of zero U_n of the division ring L, and n can be chosen in such a way that $\bar{U}_n$ is compact. From this fact it follows that from the elements of sequence (6.18) with large enough numbers it is possible to choose an element c such that for large enough p and q we have

$$(a_p - c) \in U_n, \ (a_q - c) \in U_n.$$

Because $\bar{U}_n$ is compact, then from the sequence

$$(a_1 - c), (a_2 - c), ..., (a_p - c), ... \tag{6.19}$$

it is possible to pick up a subsequence that converges to some element $b \in \bar{U}_n$. From the fact that sequence (6.18) is a Cauchy sequence immediately follows that sequence (6.19) is also a Cauchy sequence. But because its subsequence converges to the element $b \in \bar{U}_n$, the sequence (6.19) itself converges to b. From this it follows that sequence (6.18) converges to the element $b + c$.

From statement G) it follows that all three division rings that we have examined are complete (see E).)

Example 2:

Let us now give an example of a topological division ring that is not complete. Such an example is a set of all rational numbers R with a neighborhood system of zero (6.2). U_n comprises all rational numbers that satisfy the condition $|x| < \frac{1}{n}$. This is the same condition that defines the neighborhood system of zero in the field D of real numbers. In that sense the topological field R of rational numbers is a subfield of the field D of real numbers. Now if

$$a_1, a_2, ..., a_n, ... \tag{6.20}$$

is some sequence of rational numbers that satisfies the Cauchy condition, then as a sequence of elements of the field D, it converges to the number a. If a is not a rational number, then sequence (6.20) does not converge in the topological field R, and, therefore the topological field R is not complete. The topological field D that includes the topological field R is a completion of R into the complete field D in the sense that every element of D is a limit for some Cauchy sequence of the field R.

In section 6.4 we will give a different topology in the field R of rational numbers that leads to another completion of R that is different from the field of real numbers.

6.3   Uniqueness theorem

The real and complex numbers came about in mathematics as a result of a long development of the notion of number mainly due to the needs of practical applications, and partially due to the internal logic of the mathematics itself. A need to count objects gave rise to a notion of natural numbers, i.e. the positive integer numbers. Commerce and a need to assess quantities brought about the rational numbers. Developing of geometry led to discovery of the fact that the length of the diagonal of a square with the length of its side equal to one cannot be measured exactly using the rational numbers, though it can be measured by a rational number with arbitrary precision. Therefore internal development of mathematics led to the occurrence of lengths that cannot be measured by the rational numbers. Thus internal development of mathematics brought about non-rational numbers that started to be called irrational.

At the beginning there was need for only a few irrational numbers, but the requirements of logical completeness let to development of a whole set of real numbers. In particular, from a logical standpoint, it was important that any Cauchy sequence be convergent. But for the rational numbers this is not true. It is possible that a Cauchy sequence of rational numbers does not converge to a rational number, but it converges to a real number. Therefore, the field of rational numbers should be extended to the field of real numbers. The negative numbers came about mainly from internal considerations of mathematics to make possible the subtraction operation, though the negative numbers had some practical applications as well. It was possible to interpret a negative number as a debt. The complex numbers came from internal considerations of mathematics because for

121

some calculations there was need to calculate square roots of negative numbers. The appearance of complex numbers was also justified to a large extent by the fact that any polynomial with real coefficients has a root, maybe not a real, but rather, complex. Complex numbers allowed significant simplification of many calculations and caused the creation of the theory of complex functions of a complex variable, which plays an important practical and theoretical role. Therefore the complex and the real numbers are products of the historical development of mathematics. Quaternions were created in an effort to generalize the notion of number. It was an attempt to generalize the complex numbers, but it did not produce any valuable results, because the absence of commutativity does not allow development of a theory of quaternion functions. Compared with the role of complex numbers and real numbers in mathematics, the role of quaternions is minuscule.

Because the appearance of real and complex numbers in mathematics reflected a particular path of development that probably could be different, it raised a natural question: could such a different path lead to different types of numbers that are very similar to the complex and real ones, but still are different? To solve this question we have to give exact definitions of the requirements that these new objects that may play the role of numbers should comply with and to find out if there are other sets of objects that comply with these requirements. It is relatively easy to come to the conclusion that a set of objects that complies with the same requirements as numbers should be a topological space. If we apply the extra requirements of local compactness and connectedness, which is very natural to do, then it turns out that the following theorem, proven by me in 1931, holds.

Theorem 6:

Any locally compact topological division ring is either a field of real numbers, or a field of complex numbers, or a quaternions division ring.

This theorem shows in particular, that the real and complex numbers are not an accidental product of historic development, but came into being by necessity as the only objects suitable to play the role of numbers.

6.4 P-adic numbers

A set R of all rational numbers constitutes a field because of the rules of addition and multiplication that are defined there. The identity element of addition is zero, and the identity element of multiplication is the number "1". P-adic numbers arise as a result of introducing in the field R of a peculiar topology that depends on some given prime number p. An intuitive idea of this topology is realized in the perception that a rational number r is considered to be smaller the better it is divided by the given prime number p. Let us write down the number r as

$$r = \frac{a}{b}p^n. \tag{6.21}$$

Here b is a natural number not divisible by p, and a is an arbitrary integer. The number n can be positive, negative, or zero. If a is divisible by p, then the exponent n can be increased by extracting the multiplier from a. The rational number r is considered to be smaller the bigger the whole number n is. Formally it is defined by introducing a neighborhood system of zero in the field R, that complies with definition (6.3).

A) A sequence

$$U_1, U_2, ..., U_n, ... \tag{6.22}$$

that is present in definition (6.3) is defined in the following way. A neighborhood U_n comprises all numbers of type (6.21) with the given n. It immediately apparent that sequence (6.22) is decreasing, and that zero is the only common element of all sets (6.22). It turns out that sequence (6.22) complies with all conditions (6.3).

Let r_1 and r_2 be two numbers from the neighborhood U_n, so they can be written in the form

$$r_1 = \frac{a_1}{b_1} p^n, \; r_2 = \frac{a_2}{b_2} p^n.$$

We have

$$r_1 + r_2 = \frac{a_1 b_2 + b_1 a_2}{b_1 b_2} p^n.$$

From this formula it follows that

$$U_n + U_n \subset U_n.$$

Because $0 \in U_n$, then instead of this relation there is the following equality

$$U_n + U_n = U_n.$$

From this equality it follows that condition a) of definition (6.3) is satisfied.

Furthermore, we have

$$r_1 r_2 = \frac{a_1 a_2}{b_1 b_2} p^{2n}.$$

Therefore, we get

$$U_n U_n \subset U_{2n}.$$

Therefore, condition b) of definition (6.3) is satisfied. From formula (6.21) for the number r it follows that if $r \in U_n$, then $-r \in U_n$. Therefore we have

$$-U_n = U_n.$$

This means that condition c) of definition (6.3) is satisfied.

Let us prove now that condition d) of definition (6.3) is also satisfied.

If r is an arbitrary element of the set U_n, then the set $(1 + U_n)^{-1}$ comprises all elements of the type $\frac{1}{1+r}$. Let us write this fraction as

$$\frac{1}{1+r} = 1 + s.$$

From this formula we get

$$s = \frac{-a}{b + ap^n} p^n.$$

From this it follows that $s \in U_n$, and therefore

$$(1 + U_n)^{-1} \subset 1 + U_n.$$

Thus condition d) of definition (6.3) is satisfied.

Let r be an arbitrary number from R. Due to relation (6.21) the number may be written as

$$r = \frac{a}{b} p^k.$$

Here b is not divisible by p, and k is an arbitrary whole number. Multiplying the neighborhood set U_n by the number r, as it is specified by the last formula, we obviously get

$$rU_n \subset U_{n+k}.$$

From this we can see that condition e) of definition (6.3) is satisfied.

A decomposition of a whole positive number into a sum of powers of the number p is used to define p-adic numbers. This decomposition is analogous to what we use to write a negative number in the decimal form, but instead of p, the number 10 is used. The possibility of such a written form is almost obvious, but we will still formulate and prove it.

B) Each whole non-negative number x can be written in the form

$$x = x_0 + x_1 p + x_2 p^2 + ... + x_n p^n, \qquad (6.23)$$

where coefficients $x_0, x_1, ..., x_n$ are the whole numbers that satisfy the following inequalities

$$0 \le x_i \le p - 1, \; i = 0, 1, ..., n.$$

If $x = 0$ in formula (6.23) all coefficients are equal to zero and the formula is correct. If $x > 0$ we will prove the assertion using induction; specifically, we will assume that for any number $x' < x$ decomposition (6.23) takes place.

Let us consider an infinite geometric progression

$$1, p, p^2, ..., p^n, p^{n+1}, ...$$

Every natural number necessary falls into one of the intervals between neighboring members of this progression. Let us suppose that

$$p^n \le x < p^{n+1}.$$

Let us divide the number x by the number p^n. We get

$$x = x_n p^n + x'. \tag{6.24}$$

Here

$$x' < p^n, \; 0 \le x_n \le p - 1,$$

because if x_n were bigger or equal to p, then x would be bigger or equal to p^{n+1} which contradicts our supposition. Because the number $x' < p^n \le x$, then it also can be decomposed into the sum analogous to (6.23), and in that sum the power of p is not bigger then $(n-1)$. From this and from equality (6.24) follows the decomposition (6.23).

We may write every whole non-negative number of R in the form (6.23), but it is not possible to write a negative number in form (6.23). A rational number of type (6.21) also cannot always be written in form (6.23). Besides, the field R with the topology defined by A) is not complete. Taking this into account we are going to slightly generalize expression (6.23). In particular, we are going to consider infinite series of powers of

p that possibly include a finite number of negative exponents, and are going to define algebraic operations on this series.

C) Let us denote as K_0^p a set of all series of the type

$$x = \sum_{i=k}^{\infty} x_i p^i; \qquad (6.25)$$

here coefficients x_i are whole numbers that satisfy the conditions

$$0 \leq x_i \leq p - 1, \ i = k, k + 1, \dots. \qquad (6.26)$$

In the set K_0^p we will introduce operations of addition and multiplication so K_0^p will become a field, and also we will introduce a topology, so K_0^p will become a topological field. We will show that the field R with topology defined by A) is included in the topological field K_0^p, and that each element of the topological field K_0^p is a limit for the included field R.

During construction of algebraic operations in the field K_0^p we will use the operation of coefficient correction, so first let us describe this operation in general terms.

This operation is similar to the device that we use in arithmetic to make addition and multiplication of decimal numbers easy when we say "write 7, keep 2 in mind".

D) Let w be

$$w = \sum_{i=k}^{\infty} w_i p^i, \qquad (6.27)$$

where coefficients w_i may not satisfy the inequalities of type (6.26). If w_k does not satisfy the inequality

$$0 \leq w_k \leq p - 1,$$

then there exists a whole number z_k that satisfies the inequality

$$0 \le z_k \le p - 1$$

and is congruent to w_k modulo p, in other words, such that

$$w_k = z_k + u_{k+1}p.$$

Let us put this expression for w_k in sum (6.27). Then we get the following expression for w:

$$w = z_k p^k + (u_{k+1} + w_{k+1})p^{k+1} + \sum_{i=k+2}^{\infty} w_i p^i. \qquad (6.28)$$

If the following inequalities

$$0 \le u_{k+1} + w_{k+1} \le p - 1,$$

are not true then there is a number z_{k+1} that satisfies the inequality

$$0 \le z_{k+1} \le p - 1$$

and is congruent to the number $u_{k+1} + w_{k+1}$ modulo p, such that

$$u_{k+1} + w_{k+1} = z_{k+1} + u_{k+2}p.$$

Putting this expression in series (6.28) we get

$$w =$$
$$z_k p^k + z_{k+1} p^{k+1} + (u_{k+2} + w_{k+2})p^{k+2} + \sum_{i=k+3}^{\infty} w_i p^i.$$

Continuing this process further we get for w, instead of the series (6.27), the series

$$w = \sum_{i=k}^{\infty} z_i p^i$$

that is of "normal" type, which means that its coefficients satisfy the inequality

$$0 \leq z_i \leq p - 1, \; i = k, k + 1, \dots.$$

Let us now define the algebraic operations in K_0^p.
E) Let

$$y = \sum_{i=k}^{\infty} y_i p^i \tag{6.29}$$

be an arbitrary element of the set K_0^p. Here k is the same as in the series (6.25), but this does not impose any constraints, because some of the coefficients in series (6.25) and (6.29) may be equal to zero. Adding formally series (6.25) and (6.29) we get

$$x + y = \sum_{i=k}^{\infty} (x_i + y_i) p^i.$$

Applying the coefficient correction operation, which was described in D), to this series we get a series in the normal form for the sum $x + y$, i.e. for the element of K_0^p that is actual the $x + y$ sum. Therefore, the sum of $x + y$ belongs to the set K_0^p. Let us define a difference $x - y$ performing formal subtraction of series (6.29) from series (6.25). We get then

$$x - y = \sum_{i=k}^{\infty} (x_i - y_i) p^i.$$

Performing correction of coefficients of this series according to the method shown in D), we get a series of the "normal" form, that is an element of K_0^p. Multiplying series x and y as series of powers of p we get for the product xy a sum of a sequence of products where one of the multipliers in the product is a power of p. Performing coefficient correction for this series, as was described in D), we get a series of the "normal" type for the product xy, that is an element of K_0^p. So we have defined the operations of addition, subtraction, and multiplication for series of type (6.25). To complete construction of the algebraic operations in the field K_0^p we have to construct the reciprocal element. Let us do that.

F) First of all let us find the reciprocal element for the series

$$\hat{x} = x_0 + x_1 p + x_2 p^2 + ..., \tag{6.30}$$

where $x_0 \neq 0$. We will look for the reciprocal element that can be represented by the series

$$y = y_0 + y_1 p + y_2 p^2 + ... \tag{6.31}$$

Multiplying formally the series $\hat{x}$ and y we get

$$\hat{x}y = w = w_0 + w_1 p + w_2 p^2 + ..., \tag{6.32}$$

where

$$w_i = x_0 y_i + x_1 y_{i-1} + ... + x_i y_0.$$

Here series (6.31) is chosen in such a way that $\hat{x}y = 1$.

Therefore, series (6.32), after coefficient correction that is described in D), should appear like this

$$w = 1 + 0p + 0p^2 + ...$$

According to D) the coefficient correction can be written as

$$z_0 = w_0 - u_1 p;$$
$$z_i = w_i + u_i - u_{i+1} p =$$
$$= x_0 y_i + x_1 y_{i-1} + ... + x_i y_0 + u_i - u_{i+1} p,$$
$$i = 1, 2, ...$$

First of all, let us notice that because $x_0 \neq 0$, its reciprocal element from the residue class field P^p mod p exists, a number θ such that θx_0 is congruent to one modulo p.

Therefore the first of the equations we need to solve, specifically

$$z_0 = x_0 y_0 - u_1 p = 1,$$

has a solution, particularly $y_0 = \theta$. For $i > 0$ we need to solve the following equations:

$$x_0 y_i + x_1 y_{i-1} + \dots + x_i y_0 + u_i - u_{i+1} p = 0.$$

These equations can be solved one by one, in succession, with increasing i. For the equation with the number i there appears only one unknown variable y_i that has a coefficient x_0, therefore this equation can be solved with respect to y_i. For every equation we can choose a root y_i that satisfies the inequality

$$0 \leq y_i \leq p - 1.$$

Therefore, we are going to build the element y that is reciprocal to the element $\hat{x}$ (see (6.31)), with $y_0 \neq 0$. But each series of type (6.25), not equal to zero, can be written as $p^l \hat{x}$ (see (6.30)) and the reciprocal element (see (6.31)) is

$$(p^l \hat{x})^{-1} = p^{-l} y.$$

In turn each such element can be written in form (6.25), and therefore, belongs to the set K_0^p. Therefore, for every element of the set K_0^p we build the element that is reciprocal to it.

In order to prove that K_0^p, along with the operations defined on it, is a field, we have to prove that addition and multiplication are associative and distributive. This is not obvious, because after each operation we have to do coefficient correction. But I will not give any proof of these results here.

Let us introduce in the field K_0^p a topology using the neighborhood system of zero in the same way as was done in definition (6.3).

G) We will define a complete neighborhood system of zero

$$U_1, U_2, \dots U_n, \dots \tag{6.33}$$

131

in the field K_0^p by including in the neighborhood set U_n all series of type (6.25), such that $k = n$, in other words all series of the kind

$$x = x_n p^n + x_{n+1} p^{n+1} + \dots \qquad (6.34)$$

Let us write down one more series of the same kind

$$y = y_n p^n + y_{n+1} p^{n+1} + \dots$$

Each of these series can be identified by the fact that its first n coefficients are equal to zero. It turns out that for the system of sets (6.33) all conditions of definition (6.3) are satisfied.

It is clear that sequence (6.33) is a sequence of decreasing sets, and the intersection of all these sets contains zero and only zero. Formally adding elements x and y of U_n we get a series w the first n coefficients of which are equal to zero. During coefficient correction this condition will not change. Therefore we have

$$U_n + U_n \subset U_n.$$

Because U_n contains zero, then instead of the last inclusion we have an equality

$$U_n + U_n = U_n. \qquad (6.35)$$

Therefore condition a) of definition (6.3) is satisfied.

Multiplying formally series x and y we get a series of which the lower power of p will be $2n$, and all coefficients previous to that will be equal to zero. This condition can not be changed by coefficient correction, so

$$U_n U_n \subset U_{2n}.$$

Therefore, condition b) of definition (6.3) is satisfied.

Transforming the series for $-x$ (see (6.34)) we get a series the first n coefficients of which are equal to zero. This condition can not be changed by coefficient correction. Therefore, we have

$$-U_n = U_n$$

From this it follows that condition c) of definition (6.3) is satisfied.

The set $(1+U_n)^{-1}$ consists of all elements of the type $\frac{1}{1+x}$ (see (6.34)). Let us represent this element of the field K_0^p in the form

$$\frac{1}{1+x} = 1 + y.$$

From this we infer $y = -\frac{x}{1+x}$. Due to F) the element $(1+x)^{-1}$ is the element of series (6.31) that starts from the zero power of p. Because the series $-x$ starts from the nth power of p, we get $y \in U_n$, and therefore,

$$(1+U_n)^{-1} \subset 1 + U_n.$$

Therefore, condition d) of definition (6.3) is satisfied.

Let x be a series of type (6.25). In other words, it is an arbitrary element of the field K_0^p. It is clear that

$$xU_n \subset U_{n+k}.$$

Therefore, condition e) of definition (6.3) is satisfied.

Let us now include the field R of rational numbers in the field K_0^p with p-adic topology defined in R (see A)). An element r (see (6.21)) of the field R, where b is not divisible by p, can be written in the form

$$ab^{-1}p^n.$$

But element b^{-1} of the field K_0^p can be written in the form

$$b^{-1} = y_0 + y_1 p + y_2 p^2 + \ldots,$$

where $y_0 \neq 0$. Therefore the element r is included in the field K_0^p. The neighborhood system (6.22) that is present in R transforms into the neighborhood system (6.33) of the field K_0^p. Therefore we have embedded the topological field R in the topological field K_0^p, preserving its topology. For the element x, represented by series (6.25), let us define the operation $b_l(x)$ by the formula

$$b_l(x) = \sum_{i=k}^{l-1} x_i p^i.$$

It is clear that $b_l(x) \in R$.
It is also clear that

$$x - b_l(x) \in U_l.$$

From this it follows that

$$\lim_{n \to \infty} b_n(x) = x,$$

and $b_n(x) \in R$. Therefore, each element of the field K_0^p is the limit of a sequence of elements of the field R embedded in K_0^p.

Example 3:
Let us consider as an example how to write the element "-1" of the field K_0^p in form (6.25) which occurs after coefficient correction (see D)).
We have

$$-1 = w = w_0 + w_1 p + w_2 p^2 + ...,$$

where $w_0 = -1$, $w_1 = 0$, $w_2 = 0$. Correcting the coefficient w_0 we get $w_0 = z_0 + u_1 p$, where $z_0 = p - 1$; $u_1 = -1$. Furthermore, $w_1 + u_1 = -1$. Therefore, as a result of this coefficient correction we get $w_1 + u_1 = -1 = z_1 + u_2 p$, where $z_1 = p - 1$, $u_2 = -1$. Continuing this process we get a fixed series

$$-1 = \sum_{i=0}^{\infty} (p-1)p^i,$$

or

$$(p-1)(1 + p + p^2 + ...).$$

The infinite series in parenthesis converges in the topological field K_0^p, because p^i tends to zero with increasing i, and therefore, we have

$$1 + p + p^2 + ... = \frac{1}{1-p}.$$

Therefore we get

$$-1 = (p-1)\frac{1}{1-p} = -1,$$

which confirms the correctness of our calculations.

6.5 Some properties of the field K_0^p of p-adic numbers

A) Let U_n be an arbitrary set of system (6.33). Let us denote θ_n as the set of all points of K_0^p that are not included in U_n. It turns out that θ_n is a closed set.

Let us suppose that the opposite is true.

Let

$$x^1, x^2, ..., x^q, ... \tag{6.36}$$

be some sequence of points of θ_n that converges to the point $x \notin \theta_n$. Then $x \in U_n$. Because sequence (6.36) converges to x, then according to definition (6.4), that means that for a large enough q we have

$$(x^q - x) \in U_n.$$

Then considering (6.35) we have

$$x^q \in x + U_n \subset U_n.$$

Therefore, all elements of sequence (6.36), starting from some large enough number, belong to U_n and can not belong to θ_n. Therefore we arrive at a contradiction, and therefore, statement A) has been proven. ∎

Note that p-adic number x defined by series (6.25) belongs to the neighborhood set U_n if and only if $b_n(x) = 0$ (see (6.33)).

Along with p-adic number x, defined by series (6.25), let us consider an arbitrary p-adic number y that is defined by series (6.29).

B) The difference $x - y$ of p-adic numbers x and y (see (6.25) and (6.29)) belongs to the neighborhood set U_n if and only if the equality

$$b_n(x) = b_n(y)$$

holds (see (6.33)).

In order to construct the difference $x - y$ we are first going to construct a formal difference between series (6.25) and (6.29), in other words the series

$$w = \sum_{i=k}^{\infty} w_i p^i = \sum_{i=k}^{\infty} (x_i - y_i) p^i \tag{6.37}$$

Let $(x_l - y_l)$ be the first coefficient of the series that is not zero. That means $x_l - y_l$ is not congruent with zero modulo p. Therefore, the first non-zero coefficient of the series after applying the coefficient correction procedure will be z_l, and we will have

$$b_l(x) = b_l(y),$$

and l is the maximum number that satisfies this condition. From this follows the correctness of statement B).

C) In order for a sequence

$$x^1, x^2, ..., x^q, ... \tag{6.38}$$

of elements of the topological field K_0^p to converge to an element x of this field, that is, in order for the following equality to occur

$$\lim_{q \to \infty} x^q = x, \tag{6.39}$$

it is necessary and sufficient that for any natural number n there is a natural number r large enough that for $q > r$ the following equality is true

$$b_n(x^q) = b_n(x).$$

According to 6.1 B) relation (6.39) is true if and only if for any natural number n there is a large enough natural number r, such that for $q > r$

$$(x^q - x) \in U_n.$$

But according to statement B), the last inclusion takes place if and only if the following relation occurs

$$b_n(x^q) = b_n(x).$$

Therefore statement C) has been proven. ∎

D) Any neighborhood set U_v (see (6.33)) is closed and compact.

Let us first prove closeness of the set U_v. Elements of U_ν satisfy the condition

$$b_\nu(x) = 0.$$

If we assume that sequence (6.38) is completely enclosed within U_ν, then for all natural numbers n large enough, the following equality occurs

$$b_n(x^q) = b_n(x) \tag{6.40}$$

(see C)). Notice that for $v < n$, and for any element y of K_0^p, the following relation occurs

137

$$b_\nu(b_n(y)) = b_\nu(y).$$

If equality (6.40) occurs then

$$b_\nu(x) = b_\nu(b_n(x)) = b_\nu(b_n(x^q)) = b_\nu(x^q) = 0.$$

Therefore,

$$x \in U_\nu,$$

hence U_ν is closed.

Let us now prove compactness of the set U_ν. Statement 6.2 C) implies local compactness of the topological field K_0^p, and thus its completeness (see 6.2 G)).

Let M be an infinite subset of elements of U_ν. Because the first element of decomposition x into a series (6.25) may take only a finite number of values, specifically p, then it is possible to select in M an infinite set M_k with elements of M such that the x_k coefficients are all identical in series (6.25). In the same way in the set M_k we can select an infinite subset M_{k+1} of elements such that all of the elements have identical coefficients x_{k+1}. In exactly the same way we can select an infinite subset M_{k+2} in the set M_{k+1}, and so we get an infinite, decreasing sequence of subsets

$$M, M_k, M_{k+1}, M_{k+2}, ...,$$

and for two elements $x^{'}$ and $x^{''}$ that belong to M_l we have the equality

$$b_l(x^{'}) = b_l(x^{''}).$$

Let us choose now an arbitrary element x^l of M_l. We have (see C))

$$\lim_{l \to \infty} x^l = x,$$

where x is an element of the set U_ν. Therefore compactness of the set U_ν has been proven.

So statement D) has been proven. ∎

Let us now investigate in more detail the neighborhood set of zero U_ν in the field K_0^p.

E) The neighborhood set U_ν can be decomposed into a finite sum of compact, mutually non-intersecting sets of the type

$$x^\alpha + U_{\nu+l}, \tag{6.41}$$

where x^α, $\alpha = 1, ..., p^l$ are non-identical elements of the neighborhood set U_ν.

Let x be an arbitrary element of U_ν. Then

$$b_{\nu+l}(x) = x_{nu}p^{nu} + x_{\nu+1}p^{\nu+1} + \cdots + x_{\nu+l-1}p^{\nu+l-1} = x^\alpha$$

is a polynomial of p with l coefficients, and each of these coefficients can take p different values. Thus the last formula contains exactly p^l different polynomials. We will denote these polynomials as x^α. A set of all $x \in U_\nu$ for which $b_{\nu+l}(x) = x^\alpha$ represents sum (6.41). Each of sets (6.41) is closed and compact. Furthermore, if $x^\alpha \neq x^\beta$, then the corresponding sets $x^\alpha + U_{\nu+l}$ and $x^\beta + U_{\nu+l}$ have no intersection. Let us suppose that the opposite is true, that there is an element y that belongs to both these sets. Then we have

$$b_{\nu+l}(y) = x^\alpha, \quad b_{\nu+l}(y) = x^\beta,$$

But this is impossible, because $x^\alpha \neq x^\beta$. Therefore statement E) has been proven. ∎

The last statement shows that the neighborhood set U_v comprises a finite number of small pieces of type (6.41). This may hold true for the whole space K_0^p.

F) In the space K_0^p there are no connected, closed, infinite subsets.

Let us suppose that M is a connected, closed subset of K_0^p. Let y be an arbitrary element of M. Then the set $M' = M - y$

is connected and contains the zero element. Since the set M' is infinite, and the neighborhood sets U_n (see (6.33)) have only one common element, zero, then there is a large enough natural number n that the neighborhood set U_n does not contain the whole set M'.

Let us denote as M_1' the intersection $M' \cap U_n$, and M_2' as the intersection $M' \cap \theta_n$ (see A)). The sets M_1' and M_2' are closed and do not intersect. Therefore the set M' is not connected, and therefore the original set M is also not connected. Thus we have come to a contradiction, and therefore statement F) has been proven. ∎

6.6 Field of polynomials over a residue class

Let P^p be a field of a residue class modulo a given prime number p (see 5.2). The expression

$$a(t) = a_0 + a_1 t + a_2 t^2 + ... + a_n t^n, \qquad (6.42)$$

where coefficients $a_0, a_1, ..., a_n$ are modulo p remainders, is called a polynomial over the field P^p.

With polynomials of type (6.42) we can perform operations of addition, subtraction, and multiplication, executing the corresponding operations in the same way as with the elements of the field P^p.

If $b(t)$ is another polynomial of type (6.42), then the expression

$$r = \frac{a(t)}{b(t)} \qquad (6.43)$$

is called a rational expression over the field P^p.

We can add, subtract, multiply, and divide ratios of type (6.43), using the conventional rules, so the set of all such ratios constitutes a field P_t^p.

We may introduce in the field P_t^p a topology in the sense that expression (6.43) is considered to be closer to zero the better it is evenly divisible by t.

A) A topology is introduced in P_t^p using the zero neighborhood system

$$U_1, U_2, ..., U_n, ... \qquad (6.44)$$

(see (6.33)). The neighborhood set U_n includes all expressions of the type

$$r = \frac{a(t)}{b(t)}t^n, \qquad (6.45)$$

and here we suppose that the polynomial $b(t)$ is not divisible by t, that its free term $b_0 \neq 0$ in the field P^p. First of all, it is clear that sequence (6.44) is a decreasing sequence, with zero as the only common element. Besides that it turns out that all five conditions of definition (6.3) for the zero neighborhood system (6.44) are satisfied here.

If

$$r_1 = \frac{a_1(t)}{b_1(t)}t^n, \quad r_2 = \frac{a_2(t)}{b_2(t)}t^n$$

are two arbitrary elements of the neighborhood set U_n, then it is clear that their sum is an element of U_n. Taking into account that $0 \in U_n$, we get the following equality

$$U_n + U_n = U_n.$$

So condition a) of definition (6.3) is satisfied.
Furthermore, we have

$$r_1 r_2 = \frac{a_1(t)a_2(t)}{b_1(t)b_2(t)}t^{2n}.$$

The denominator of the ratio $b_1(t)b_2(t)$ is obviously not divisible by t. Hence, we have

$$U_n U_n \subset U_{2n}.$$

141

Therefore, condition b) of definition (6.3) is satisfied.
Furthermore, we have (see (6.45))

$$-r = -\frac{a(t)}{b(t)}t^n.$$

From this we get

$$-U_n = U_n,$$

and condition c) of definition (6.3) is satisfied.

Element (6.45) is an arbitrary element of the neighborhood set U_n. Therefore an arbitrary element of the set $(1 + U_n)^{-1}$ may be written as $(1 + r)^{-1}$. If we suppose that

$$\frac{1}{1+r} = 1 + s,$$

then for s we get the expression

$$s = -\frac{a(t)}{b(t) + a(t)t^n}t^n.$$

Therefore,

$$(1 + U_n)^{-1} \subset 1 + U_n,$$

so condition d) of definition (6.3) is satisfied.

Let

$$s = \frac{a(t)}{b(t)}t^k$$

be an arbitrary element of the field P_t^p. Then we obviously get

$$sU_n \subset U_{n+k},$$

so condition e) of definition (6.3) is satisfied.

So sequence (6.44) is a neighborhood system of zero in the field P_t^p, and thus P_t^p becomes a topological field. It turns out that this field is not locally compact though. In order to include it in a compact complete field, we will consider the series of t with coefficients of P_t^p, including maybe a finite number of negative exponents of t.

B) A set of all series of the type

$$x = \sum_{i=k}^{\infty} x_i t^i, \tag{6.46}$$

where coefficients x_i are elements of the field of residue class P^p, and k is a nonzero integer, we denote as K_t^p.

The operations of addition, subtraction and multiplication can be naturally defined in the set K_t^p. To construct a reciprocal element, let us first find the reciprocal element of the polynomial

$$\hat{x} = x_0 + x_1 t + x_2 t^2 + ..., \tag{6.47}$$

where $x_0 \neq 0$. Let us denote the reciprocal element as y and write it down as

$$y = y_0 + y_1 t + y_2 t^2 + ... \tag{6.48}$$

Let us denote the product of $\hat{x}y$ as w. Then we will have

$$w = \sum_{i=0}^{\infty} w_i t^i,$$

where

$$w_i = x_0 y_i + x_1 y_{i-1} + ... + x_i y_0.$$

In order for y to be the reciprocal of $\hat{x}$, it is sufficient to satisfy the following conditions

$$w_0 = 1, \quad w_i = 0, \quad i = 1, 2, \dots \qquad (6.49)$$

The first of these equations gives us the relation

$$x_0 y_0 = 1.$$

This equation is solvable in the field P^p because $x_0 \neq 0$. The next equation in sequence (6.49) is

$$x_0 y_1 + x_1 y_0 = 0.$$

We should solve it for y_1, which is possible, because the coefficient of y_1 is not zero. Every following equation of the sequence (6.49) contains only one new unknown element that comes with a non-zero coefficient. Therefore, the element y (see (6.48)) that is reciprocal to $\hat{x}$ can be calculated, and $y_0 \neq 0$. An arbitrary series (6.46), not equal to zero, can be written in the form

$$x = t^l \hat{x}$$

(see (6.47)), and we get

$$x^{-1} = t^{-l} y,$$

thus the reciprocal element for $x \neq 0$ has been constructed. Hence K_t^p is a field.

Let us set up a topology in this field following the method shown in definition (6.3).

C) Let us define a neighborhood system of zero of the field K_t^p as a sequence

$$U_1, U_2, \dots, U_n, \dots \qquad (6.50)$$

where the neighborhood set U_n comprises all series (6.46) in which $k = n$.

First of all it is clear that sequence (6.50) is a decreasing sequence, and the only common element for all its members is zero. Further, it turns out that for this neighborhood system of zero all conditions of definition (6.3) are satisfied.

It is easy to check that the following relations occur

$$U_n + U_n = U_n; \quad U_n U_n = U_{2n}; \quad -U_n = U_n.$$

From this it follows that conditions a), b) and c) of definition (6.3) are satisfied.

Let us check condition d). The set $(1 + U_n)^{-1}$ comprises all elements of the type $(1 + x)^{-1}$, where in a series for x (6.46) $k = n$, $n \geq 1$. Therefore, there exists the reciprocal element for $(1 + x)$ (see B)). Its series starts with 1 and then following exponents of t starting from the n-th one. As a result we get

$$(1 + U_n)^{-1} \subset (1 + U_n),$$

and, hence, condition d) of definition (6.3) is satisfied.

If x (see (6.46)) is an arbitrary element of the field K_t^p, then it is clear that

$$xU_n \subset U_{n+k}.$$

Therefore condition e) of definition (6.3) is satisfied.

So the neighborhood system of zero (6.50) defines a topology in the field K_t^p. Let us embed P_t^p in the field K_t^p.

D) Each element r of the field P_t^p may be written in the form (see (6.43))

$$r = \frac{a(t)}{b(t)} t^n.$$

Because the polynomial $b(t)$ is not divisible by t, then in the field K_t^p it has the reciprocal element $b^{-1}(t)$ (see B)), and the element r can be written in the form

$$r = a(t)b^{-1}(t)t^n,$$

and is included in the field K_t^p. Therefore, the topological field P_t^p is included in the topological field K_t^p, and each neighborhood set U_n of series (6.44) is a subset of the neighborhood set U_n of series (6.50).

Thus, the topological field P_t^p is embedded in the topological field K_t^p with preservation of its topology.

E) Let us correspond for each sum x (see (6.46)) a finite sum

$$b_l(x) = \sum_{i=k}^{l-1} x_i t^i.$$

First of all, it is clear that the condition $b_l(x) = 0$ is equivalent to the condition $x \in U_l$. From that it follows that the relation

$$\lim_{q \to \infty} x^q = x$$

occurs if and only if for any arbitrary natural number n there can be found a natural number r, such that for $q > r$ we have:

$$b_n(x^q) = b_n(x). \tag{6.51}$$

From this condition it immediately follows that

$$\lim_{q \to \infty} b_n(x) = x,$$

but $b_n(x)$ is an element of the field P_t^p. So the closure of the field P_t^p that is embedded in the field K_t^p, is the whole field K_t^p.

F) Let us construct a one-to-one mapping f of the space K_t^p onto the space K_0^p of p-adic numbers. In order to construct such a mapping f let us replace in series (6.46) every remainder x_i by a number that represents x_i, which satisfies the inequalities $0 \le x_i \le p - 1$, and replace the letter t by a prime number p. The series (6.46) becomes a p-adic number $f(x)$. This mapping does not preserve the algebraic operations, because there is a coefficient correction in the field of p-adic numbers. But it preserves convergence (see (6.51), (6.40)). Specifically, if

$$\lim_{q \to \infty} x^q = x, \tag{6.52}$$

then

$$\lim_{q \to \infty} f(x^q) = f(x),$$

and vice versa, from the last relation follows (6.52). From the statement F) it follows that the field K_t^p does not contain any connected sets.

6.7   About a structure of not connected locally compact topological division rings

Although the locally compact connected topological spaces are thoroughly investigated, regarding the non-connected compact spaces I can provide only Kovalsky's theorem that he proved in 1953.

Theorem 7:

A locally compact non-connected division ring L is necessarily disconnected everywhere, that is, it does not contain any connected subsets, and there can be two mutually exclusive cases:

  a) *the division ring L characteristic is zero, and in that case it contains the field $K_0^p = K$ of p-adic numbers;*

  b) *the division ring L characteristic is p, and in that case it contains the field $K_t^p = K$ of series of some t.*

In both these cases elements of the field K are commutative in multiplication with elements of the division ring L, and there is a finite linear basis of the division ring L over the field K. Specifically, there is a system of elements

147

$$l_0 = e, l_1, ..., l_\nu,$$

such that each element $x \in L$ can be written as

$$x = x_0 l_0 + x_1 l_1 + ... + x_\nu l_\nu,$$

where the coefficients $x_i \in K$.

6.8 Summary

In this book we examined systems of elements with defined algebraic operations, and the limit operation, that are logically conceivable extensions of numbers. In particular, applying to such a system some very general constraints we came to the conclusion that there are no other logical possibilities satisfactory to mathematics for constructing objects that are similar to the real and complex numbers besides the real and complex numbers themselves. It shows that the real and complex numbers appeared in mathematics not as result of whimsical historic development, but as the only possible objects that satisfy the conditions which are naturally assumed to be applicable to numbers.

Made in the USA
Las Vegas, NV
13 November 2021